"*Transforming Socio-Natures* is a landmark text for understanding Turkey's socio-environmental challenges. This interdisciplinary collection engages the reader in a growing understanding of the ecological costs of neoliberal authoritarianism in Turkey. Ultimately, the reader emerges with critical tools for building a new historical-geographical framework for environmental thought and practice."
    –**Anna J. Secor**, Professor of Human Geography, Durham University, UK

"Smart, timely, incisive—this book is a wonderful collection representing some of the best new work from emerging scholars on issues of changing socio-natures, and environmental politics in contemporary Turkey. It is precisely the sort of book I would have loved to have had years ago, but at least its time has come."
    –**Leila M. Harris**, Professor, Institute for Resources Environment and
                Sustainability and the Institute for Gender, Race, Sexuality
                and Social Justice, University of British Columbia, Canada

"Spanning the whole Republican era and a wide range of environmental issues, this book fills a lacuna in studies of Turkey. It provides the reader with a unique vantage point for understanding the trajectories of and contemporary consequences of the vast economic, demographic and political transformations in the country."
                –**Ståle Knudsen**, Professor of Social Anthropology,
                                University of Bergen, Norway

Transforming Socio-Natures in Turkey

This book is an exploration of the environmental makings and contested historical trajectories of environmental change in Turkey. Despite the recent proliferation of studies on the political economy of environmental change and urban transformation, until now there has not been a sufficiently complete treatment of Turkey's troubled environments, which live on the edge both geographically (between Europe and Middle East) and politically (between democracy and totalitarianism).

The contributors to *Transforming Socio-Natures in Turkey* use the toolbox of environmental humanities to explore the main political, cultural and historical factors relating to the country's socio-environmental problems. This leads not only to a better grounding of some of the historical and contemporary debates on the environment in Turkey, but also a deeper understanding of the multiplicity of framings around more-than-human interactions in the country in a time of authoritarian populism.

This book will be of interest not only to students of Turkey from a variety of social science and humanities disciplines but also contribute to the larger debates on environmental change and developmentalism in the context of a global populist turn.

Onur İnal is a researcher based in the Near Eastern Studies Department, University of Vienna, Austria.

Ethemcan Turhan is a researcher based in the Environmental Humanities Laboratory, Division of History of Science, Technology and Environment, KTH (Royal Institute of Technology), Sweden.

Routledge Environmental Humanities

Series editors: Scott Slovic (*University of Idaho, USA*),
Joni Adamson (*Arizona State University, USA*) and
Yuki Masami (*Kanazawa University, Japan*)

The *Routledge Environmental Humanities* series is an original and inspiring venture recognising that today's world agricultural and water crises, ocean pollution and resource depletion, global warming from greenhouse gases, urban sprawl, over-population, food insecurity and environmental justice are all *crises of culture*.

The reality of understanding and finding adaptive solutions to our present and future environmental challenges has shifted the epicenter of environmental studies away from an exclusively scientific and technological framework to one that depends on the human-focused disciplines and ideas of the humanities and allied social sciences.

We thus welcome book proposals from all humanities and social sciences disciplines for an inclusive and interdisciplinary series. We favour manuscripts aimed at an international readership and written in a lively and accessible style. The readership comprises scholars and students from the humanities and social sciences and thoughtful readers concerned about the human dimensions of environmental change.

Transforming Socio-Natures in Turkey

Landscapes, State and Environmental Movements

Edited by
Onur İnal and Ethemcan Turhan

First published 2020
by Routledge
2 Park Square, Milton Park, Abingdon, Oxon OX14 4RN

and by Routledge
605 Third Avenue, New York, NY 10017

First issued in paperback 2021

Routledge is an imprint of the Taylor & Francis Group, an informa business

British Library Cataloguing-in-Publication Data
A catalogue record for this book is available from the British Library

Library of Congress Cataloging-in-Publication Data
A catalog record for this book has been requested

ISBN 13: 978-0-367-78574-1 (pbk)
ISBN 13: 978-1-138-36769-2 (hbk)

Typeset in Goudy
by Apex CoVantage LLC

Contents

Illustrations

Figures

Tables

Notes on contributors

Eda Acara is an assistant professor of geography in İzmir Bakırçay University, Turkey. She received her B.A. in Sociology from the Middle East Technical University, her M.A. in Gender and Women Studies from St. Mary's University and Mount Saint Vincent University and her Ph.D. from Department of Geography, Queen's University, Canada. She is a feminist environmental geographer problematizing the effects of neo-liberalization within the context of material mobility across rural-urban nexus, including water pollution and food. Her most recent work on the political ecology of Ergene River appeared in *Annals of the American Association of Geographers*.

Cem İskender Aydın is 2018/2019 Mercator-IPC fellow on climate change in Istanbul Policy Center, Sabancı University, Turkey. He received his B.A. and M.A. in economics from Boğaziçi University, another M.A. in environmental economics from Toulouse School of Economics, France and his Ph.D. in ecological economics from the Université de Versailles Saint Quentin-en-Yvelines, France. As an ecological economist working on energy and climate policy, environmental justice and mapping environmental conflicts, he has published in *Frontiers in Energy Research*, *Capitalism Nature Socialism* and *New Perspectives on Turkey*.

Üstün Bilgen-Reinart is a Turkish-Canadian writer, journalist and broadcaster. She is the author of three notable books; the first two on the social and environmental dislocations associated with development in Canada and western Turkey, respectively on the ordeals experienced by the relocated Aboriginal peoples of Canada, the Sayisi Dene First Nation in Tadoule Lake (Manitoba) and then by Bergama villagers of Turkey's Aegean Region campaigning against gold mining in their land. Her latest book is the autobiographical *Porcelain Moon and Pomegranates: A Woman's Trek through Turkey* (Dundurn Press, 2008).

Seçil Binboğa is a Ph.D. candidate in the Architectural History and Theory Program at the University of Michigan, USA. She is interested in relations between nature, space, and technology with a particular focus on the urban and environmental histories of the modern Middle East. Her dissertation research explores the politics of spatial planning and infrastructural development in Cold War Turkey.

Onur İnal is a postdoctoral researcher at the Near Eastern Studies Department, University of Vienna and founder of Network for the Environmental History of Turkey (NEHT). He received his Ph.D. in History from the University of Arizona. Previously, he was the managing director of the TurkeiEuropaZentrum (TEZ) of the University of Hamburg. His research focuses on the urban and environmental histories of the late Ottoman Empire and early Republican Turkey. He is the co-editor of *Seeds of Power: Explorations in Ottoman Environmental History* (White Horse Press, 2019).

Ekin Kurtiç is a Ph.D. candidate in Social Anthropology and Middle Eastern Studies Program at Harvard University, USA, and a graduate student associate at the Weatherhead Center for International Affairs. She holds a B.A. degree in Political Science and International Relations from Boğaziçi University. Her research interests lie in the areas of anthropology of infrastructure, environmental history, political ecology, and political anthropology with a particular focus on the water-forest relationship, rural transformations, and the making of the state and expertise in Turkey.

Ekin Gündüz Özdemirci is a researcher in film studies. She received her Ph.D. from Marmara University in 2012. Her main research interest is the question of filmic representations, which she recently combined with environmental studies. She pursued her work on the analysis of ecological representations in Turkish films during a fellowship at the Rachel Carson Center for Environment and Society, LMU Munich between 2017 and 2018. Previously, she was a visiting scholar at the Brunel University, London, between 2014 and 2015 where she conducted a research on environmental sustainability in the British film industry. Besides her academic work, she participated in the Applied Ecovillage Living Training in Findhorn Ecovillage and has a permaculture design certificate.

Begüm Özkaynak is a professor of economics at the Department of Economics, Boğaziçi University, Turkey. She holds a B.A. and M.A. in Economics from Boğaziçi University, an M.Phil. in Economics from University of Manchester and a Ph.D. in Ecological Economics and Environmental Management from ICTA, Universitat Autònoma de Barcelona. She was a coordinating lead author on Scenarios and Transformative Change in UNEP-Global Environmental Outlook Report 5 and a lead author in UNEP-Global Environmental Outlook Report 6 on Outlooks and Emerging Issues.

Dale J. Stahl is Assistant Professor of History at the University of Colorado Denver, USA. He received his Ph.D. from Columbia University in 2014. An environmental historian specializing in the modern Middle East, his research focuses on state formation and governance, environmental resource management, international relations, and economic development. He is currently writing an environmental history of the Tigris-Euphrates River Basin in the twentieth century. His research has been supported by the New York University Abu Dhabi Institute, the Institute of Turkish Studies, the Whiting Foundation, and the American Academic Research Institute in Iraq.

Gül Tuçaltan received her Ph.D. from the Department of Human Geography and Planning at Utrecht University, the Netherlands in 2017. Her doctoral research focused on the urban political ecologies of waste. Prior to her doctoral studies, she worked in academic and professional institutions such as Utrecht University, Middle East Technical University, TU Darmstadt, REC Turkey and UNIDO. Her main research interests are urban and infrastructure policies, politics and governance, urban political ecology, the production of socioecological injustice, processes of (urban) social exclusion and inclusion, and the politics of waste and waste governance.

Ethemcan Turhan is a postdoctoral researcher at the Environmental Humanities Laboratory, Division of History of Science, Technology and Environment, KTH (Royal Institute of Technology), Sweden. He received his B.Sc. in environmental engineering from Middle East Technical University and M.A. and Ph.D. degrees from ICTA, Universitat Autònoma de Barcelona. Previously, he was 2014/2015 Mercator-IPC fellow on climate change in Istanbul Policy Center, Sabancı University. Situated in political ecology tradition, his scholarly work appeared on academic outlets such as *Applied Energy, Sustainability Science, Capitalism Nature Socialism, New Perspectives on Turkey, Turkish Studies, Energy Research and Social Science, WIREs Climate Change, Ecological Economics, Journal of Political Ecology* and *Global Environmental Change.*

Zehra Taşdemir Yaşın received her Ph.D. in Sociology from State University of New York at Binghamton in 2014 and currently works as a development specialist at Ankara Development Agency under the coordination of the Ministry of Industry and Technology, Turkey. Her research interests are historical sociology, environmental sociology, political ecology and social theory with a specific focus on social-ecology of oil extraction, agro-pastoral transformation, historical sociology of nation-state formation and environmental resistance in the Middle East and Turkey. Her works appeared in outlets such as the *Journal of Historical Sociology* and *Journal of Peasant Studies.*

Sezai Ozan Zeybek is a Freudenberg Stiftung Fellow at the Forum Transregionale Studien in Berlin, Germany. He received his Ph.D. in Geography from The Open University, UK. He worked as an Associate Professor at Istanbul Bilgi University, Turkey. His work focuses on more-than-human collectives and ecologies, post-colonial formations of modernity, state violence, militarism and different constructions of manhood. He also writes children's stories.

Acknowledgments

This edited volume is the fruit of collective labor of many people. We gratefully acknowledge the scholarly support of Ece Aksoy, Michael Christopher Low, Kyle Evered, Zafer Yenal, Ayfer Bartu Candan, Erdem Evren, Sinan Erensü, Hayriye Özen, Alevgül Şorman, Marco Armiero, David Ingram, Rana İzci Connelly, Osman Balaban and Baran Alp Uncu for their detailed comments on individual chapters. Ethemcan Turhan also wishes to acknowledge the *ACKNOWL-EJ* (ISSC, Transformations to Sustainability Programme) and *Occupy Climate Change!* (FORMAS) projects. Personal thanks are especially due to the authors, who, despite their busy schedules, diligently prepared their contributions, welcomed our critique and revised and re-revised their texts. All this happened in a backdrop where scholars in Turkey (exemplified by the now well-known case of *Academics for Peace*, https://barisicinakademisyenler.net/English) face increasing risks due to strangling academic freedoms and freedom of speech. Leila Walker from Routledge deserves recognition for her encouragement and push to keep us on track. A special and heartfelt gratitude goes to our families (Alaz, Sarah, Leyla, Çınar, Fidel and Ekin) who did the heavy-lifting of bearing with us for the time we spent away working on this book. This one is dedicated to them.

1 Socio-natures on the edge

Landscapes, state and movements in Turkey

Ethemcan Turhan and Onur İnal

Standing in front of a full riot gear brigade of *jandarma* (rural military police) and an excavation machine, *Havva Ana* (Rabia Özcan, a woman in her late 50s known by the pseudonym, Mother Havva) was claiming with fervor:

> *"Who is the state? I am the people, who is the state? The state exists thanks to us, the people. The roads between these highlands will not be connected. We most certainly do not want it. The governor calls us marauders. We have been living here since birth. Who is the governor, who is the district governor? I am the people!"*[1]

This elderly rural woman's outrage in July 2015 on the rain-soaked, lush green highlands of the Eastern Black Sea coast of Turkey summarizes the uncanny transformation of local environments, state-capital relations and the landscapes of authoritarian neoliberalism, hydropower politics, environmental opposition, and mass tourism investments disguised as development in a nutshell. Undertaken by a construction-energy conglomerate which moved up the ranks of 'usual suspects' in the past decade due to its very close ties with the Erdoğan government, the so-called Green Road project aiming at building a 2,600 km highway through the pristine highland provinces of the region was the target of Havva Ana's rage.[2] The main rationale behind this project was to provide infrastructure for highland tourism, thereby transforming a region once best known for its state-supported tea and tobacco production into a mass tourism hotspot (Alaeddinoğlu and Şeremet, 2016). The rural transformation and dispossession in the region, in part driven by IMF-imposed liberalization of the agriculture and energy sectors after 2001, yielded a strong grassroots environmentalist movement, particularly against small hydropower projects (Sayan, 2017). The deterioration of rural livelihoods and consequent outmigration were quickly entrenched into the fabric of these highlands, which carry memories of religious conversions, multiple ethno-linguistic identities and state-building (Simonian, 2007), while also demonstrating the strong tradition of left-wing politics and resistance to state power in Turkey (Biryol, 2012).

Among many others, the story of *Havva Ana* and the Green Road urges us to explore the transformation of socio-natures from a variety of multi-scalar and trans-disciplinary perspectives. Turkey is one of the missing pieces of the global environmental humanities puzzle. Likewise, environmental humanities is the missing piece in the field of Turkish studies. With its attention to "thick inter-weavings of human cultures, histories, values, imaginaries and ways of life with a dynamic more-than-human-world" (van Dooren, 2018: 418), environmental humanities promises to fill a significant gap in the field of Turkish studies. This is what makes this book timely, well positioned and valuable: introducing Turkey, a country on the edge, into environmental humanities scholarship, while bringing environmental humanities to the attention of scholars of Turkey dealing with the social, historical, economic and cultural dimensions of environmental change along the society-nature continuum. Turkey is a burgeoning field of study for environmental humanities/social sciences not only due to its ambiguous geopo-litical positioning, but also because of its troubled relationship with development and modernization (Adaman and Arsel, 2005; Adalet, 2018).

This book is the product of a scholarly collaboration extending into environ-mental humanities by benefiting mainly (but not exclusively) from the strengths of trans-disciplinary approaches in particular from environmental history and polit-ical ecology. Therefore, our broader attention to the socio-natures of a country on the edge is no coincidence. Following Swyngedouw (1999), we refer to socio-nature as a historical-geographical process that is at once malleable, transformable and transgressive (446–447). Describing 'social nature' from a critical standpoint, Castree (2001) suggests that nature is intrinsically social across different scales in different ways, and one needs to focus on revealing the perpetuation of power and inequality to avoid abstracting nature from its inextricable social qualities. In this continuum, the environment often refers to a nonsocial and nonhuman nature, although the history of environment as an idea has taken different turns in the past century (Warde et al., 2018). Our conceptualization of socio-natures here follows the material and discursive co-production of nature and society in Tur-key through historically situated processes. This includes attention to the state-society relations "sedimented in relation to past histories and geographies and also recast in relation to recent development and environmental changes" (Har-ris, 2012: 35). We therefore start from *"the actually existing socionatural conditions,"* which are "always the result of intricate transformations of preexisting configu-rations that are themselves inherently natural and social" (Swyngedouw, 1999: 445, our emphasis). Thus, our attention to the production of uneven landscapes, socio-ecological hierarchies and resistance to them provides fresh contributions to the environmental social sciences and humanities literature on Turkey. Conse-quently, throughout this book the individual chapters trace the transformation of socio-natures in the Republic of Turkey through "multiple narratives that relate material practices, representational visions and symbolic expressions" (Swynge-douw, 2015: 21).

While the environmental histories of the Ottoman Empire have finally received well-deserved attention (İnal and Köse, 2019), critically informed historical-geographical research has only recently begun to explore the interplay between the environment and the country's tumultuous political, economic and histori-cal trajectory during the Republican period. Despite the recent proliferation of studies of the political economy of environmental change, rural dispossession and urban transformation in particular, an elaborate exploration on the environmental meaning makings and contested historical trajectories of environmental change in Turkey still require attention. Prominent contributions to this literature succinctly depict perspectives mostly from political science, ecological economics, political sociology and urban studies in critically reflecting on and contributing to the debates related to the politics and economics of the Erdoğan regime (Aksu et al., 2016; Özbay et al., 2016; Adaman et al., 2017; Göçek, 2018). In particular, the Gezi Park protests of 2013 sparked a renewed scholarly interest in the long-brewing environmental conflicts in the country (Özkaynak et al., 2015; Aksu and Korkut, 2017).

Contributions from human geography and critical environmental politics have been instrumental in exposing the socio-ecological impact of Recep Tayyip Erdoğan's strong commitment to socio-environmentally destructive megaprojects, transportation and infrastructural investments, such as the third bridge over the Bosporus, a gigantic airport in the middle of Istanbul's last standing northern forests, and a Suez-type canal to connect the Marmara Sea to the Black Sea. Add Turkey's construction and imported energy-driven economic growth obsession, and we have a clearer picture. Yet, we believe that this is not enough. A more complete treatment of Turkey's troubled environments on the edge calls for employing the toolbox of environmental humanities and exploring the main political, cultural and historical factors related to socio-environmental problems (including, but not limited to climate change, see Turhan, 2017). This helps us not only to better ground some of the historical and contemporary debates on the *actually existing socio-natures* in Turkey, but also to understand the multiplicity of framings and narratives around human-nonhuman interactions in the country in a time of authoritarian populism. Moreover, we believe that Turkey's unique position of living *on the edge* both geographically (between Europe and Middle East) and politically (between democracy and totalitarianism) is an interesting case to explore from an environmental humanities perspective. This is what this book tries to achieve.

As with many similar projects, the initial seeds of this book were sown during an academic workshop held at TürkeiEuropaZentrum (TEZ) of the University of Hamburg in October 2017, which also served as the inaugural meeting of the Network for the Study of Environmental History of Turkey (NEHT, www.envhistturkey.com). The project was later expanded in scope by inviting carefully solicited chapters from an emerging cohort of early career researchers in the field. Hailing from disciplines as diverse as history, sociology, urban studies, anthropology, architecture, political ecology, geography and film studies, these early career researchers transgressed the boundaries of their own disciplines by situating their contributions to this book within the overarching framework of the environmental humanities. Expanding on the existing works on the roots and directions of environmental change and analyzing variegated relationships between the state and society, between power and nature, between environmental change and economic change, between the core and the periphery, and between the urban and the rural, this book offers a holistically novel perspective on the landscapes, state-capital axis and environmental movements in Turkey. One shortcoming of this collection is its lack of further attention to ecocritical approaches to the Turkish literary tradition. However, on this we happily refer the attentive reader to a recent addition to ecocritical literature that successfully covers this lacuna (Gürses and Ertuna Howison, 2019). Consequently, the chapters here, we hope, will be of interest not only to students of Turkey from a variety of social science and humanities disciplines, but also contribute to the larger debates on environmental change in the context of the global authoritarian populist turn. On this latter point, we also believe that the contributions in this book are fitting with the growing scholarly interest in social, cultural, economic and political struggles over the environment (for recent studies on environmental governance under authoritarian populism, see McCarthy, 2019; Brain and Pál, 2019). After all, as C. S. Lewis reminds us in *The Abolition of Man* (1943: 55), "What we call man's power over nature turns out to be a power exercised by some men over other men with nature as its instrument." Merging a critical diagnosis and a normative prognosis for a country on the edge, this book is our collective tribute to *Havva Ana*'s plea.

Overview of the book

The book is organized into three sections that, respectively, cover transforming landscapes, transforming state-capital relations and transforming environmental movements. The first section focuses on the production of transforming landscapes on the edge.

In Chapter 2, Binboğa traces the changing meanings of nature informing the aspirations for agricultural development and industrialization between 1930 and 1960. In an attempt to demonstrate how expert knowledge contributed to the making of an experimental nature, her chapter focuses on scientific soil survey reports produced by the sugar technologist, Cornelis van Dillewijn, the soil specialists, Willem van Liere and Harvey Oakes, and the agronomist, Leo Placide Hebert, all of took up roles in Turkey's modernization drive. This chapter helps us to reconsider entwined conceptions of land as experimental nature that is continuously produced and re-produced.

In Chapter 3, Stahl takes us through a journey to a techno-political frontier: the Keban Dam project. Proving to be one of the largest civil engineering projects in the world in the 1960s–1970s, the making of the Keban Dam not only relied on the idea of eastern Anatolia as a frontier region, but also taking the dam itself as a frontier to be crossed in making the imagining and engineering of a young nation possible. The chapter focuses on the ideas, influences and changes brought to light by multiple actors in this period. Using both historical and literary sources, this chapter successfully narrates the enfolding of the historical frontier into a broader techno-political framing of southeastern Anatolia.

In Chapter 4, Acara moves across the country to provide a detailed account of the former imperial frontier in Turkey's European province, Thrace. By exploring the meanings and representations of unruly places in this region where socio-natures have been interwoven with nation-building practices since the demise of the Ottoman Empire, this chapter employs a discourse analysis of the micro-histories of the Alpullu Sugar Factory and the Ergene River. Through a well-crafted mix of human and more-than-human histories, this chapter provides a lively account of the social and discursive constructions of Turkish territory and territoriality in the period from 1920 to 1940.

The second section of the book focuses on the transforming state and capital relations on the edge.

In Chapter 5, Zeybek focuses on the uncanny assemblages of humans and non-humans in re-ordering the spatial arrangements of animal husbandry and industrial meat production against a backdrop of ethnic strife in the Kurdish provinces of southeastern Turkey. Contributing to the literature on the capitalization of agriculture and the securitization of rural livelihoods, this chapter focuses on the notion of animal population control as key for controlling human populations from the state's vantage point. The chapter advances the notion that capital accumulation by and large depends on disciplining, monitoring and supervising human and nonhuman bodies, enabled by new constellations of relations between trees, soil, people, animals and water.

In Chapter 6, Kurtiç provides a detailed reading of the history of dams in Turkey over sediments and the relations between forest landscapes, the state's interest in taming natures through technical intervention and natural resource rehabilitation projects in the Çoruh River Valley of northeastern Turkey. Using historical and ethnographic methods, this chapter narrates a "sedimented history" of nation-state making, developmentalism and rising authoritarianism through its attention to the state's moves to know, measure and control rivers. In doing so, it chronicles the blending of rural transformation and depopulation, the rise of socio-environmental movements and the state's green rhetoric about the water-forest nexus.

In Chapter 7, Tuçaltan takes us through the socio-spatial, material and economic entanglements between the poverty, waste and urban growth in the country's capitol, Ankara, between the 1930s and 1990s. Focusing on the production of informal waste regimes, and the materialities, actors, diverse representations and power relations therein, her chapter shows how waste moves between being matter out of place and a source of value creation. By doing so, Tuçaltan's contribution provides a detailed account of socio-historical struggles waged over waste in tandem with macro-economic, political and institutional transformations in Turkey.

The last section in the book focuses on transforming movements on the edge, referring to movements in the broadest sense of the word.

In Chapter 8, Yaşın offers an exploration of two emblematic anti-mining struggles (Bergama and Cerattepe) in unraveling the geographically and historically specific relationships between rampant capitalist development and rapid socio-ecological destruction. By contextualizing the rise of environmental movements in these two cases, the chapter shows how languages of resistance have been gradually transformed to reflect the historical context of their emergence. The contribution underlines how the transformation of state-society-nature relations in Turkey manifests as the dissolution of locally specific human-nature bonds as part of the global expansion of extractive commodity frontiers.

In Chapter 9, Turhan, Özkaynak and Aydın revisit one of the first nationally debated environmental justice victories against coal-fired power plant investments in the Aegean region of western Turkey. Providing a micro-historical account of the Aliağa anti-coal movement, the chapter focuses on the changing nature and shifting contours of environmental mobilizations in the country in the past three decades. Through a discussion of legal struggles and the deteriorating rule of law in the country, this contribution unravels the continuities, ruptures and tipping points between the state and environmental movements at a historical junction when anti-coal struggles are being reborn in the name of climate justice.

In Chapter 10, Özdemirci scrutinizes the transforming ideas of nature in moving pictures with a focus on *New Turkish Cinema*. Through an ecocritical reading of realist art house cinema, the chapter suggests that changing depictions of nature in the past decades imply an intuitive approach to nature and hint at a recognition of nature's intrinsic value. By analyzing the nature-based aesthetic in cinema, the chapter sheds light on the quest for belonging to enchanted nature in a time when Turkish society is increasingly polarized.

To conclude the book, Bilgen-Reinart offers a *tour d'horizon* in her Epilogue regarding the transformation of socio-natures in Turkey in the past three decades by connecting the dots between the local and the global.

Notes

1 Yeşil Yol'a Havva Ana isyanı: "Halkım ben!", www.youtube.com/watch?v=bUpG2fL7Cic (Accessed on 25/03/2019).
2 For analyses of embodied and transversal female political subjectivities and visual discourses of Havva Ana and other rural women in grassroots environmental movements in Turkey, see İrhan, 2016; Yaka, 2017.

References

Adalet, B. (2018). *Hotels and Highways: The Construction of Modernization Theory in Cold War Turkey*. Stanford, CA: Stanford University Press.

Adaman, F., Akbulut, B., and Arsel, M. (eds.). (2017). *Neoliberal Turkey and Its Discontents: Economic Policy and the Environment Under Erdoğan*, 1st ed. London; New York: I. B. Tauris.

Adaman, F. and Arsel, M. (2005). *Environmentalism in Turkey: Between Democracy and Development*. London; New York: Routledge.

Aksu, C., Erensü, S., and Evren, E. (2016). *Sudan Sebepler: Türkiye'de Neoliberal Su-Enerji Politikaları ve Direnişler*. Istanbul: İletişim Kitapları.

Aksu, C. and Korkut, R. (2017). *Ekoloji Almanağı: 2005–2016*. Istanbul: Yeni İnsan Yayınevi.

Alaeddinoğlu, F. and Şeremet, M. (2016). Nature-Based Tourism in Turkey: The Yayla in Turkey's Eastern Black Sea Region. In I. Egresi, ed., *Alternative Tourism in Turkey: Role, Potential Development and Sustainability*, 1st ed. Heidelberg: Springer, pp. 71–86.

Biryol, U. (2012). *Karardı Karadeniz*. Istanbul: İletişim.

Brain, S. and Pál, V. (eds). (2019). *Environmentalism Under Authoritarian Regimes. Myth, Propaganda, Reality*, 1st ed. London; New York: Routledge.

Castree, N. (2001). Socializing Nature: Theory, Practice, and Politics, In N. Castree and B. Braun, eds, *Social Nature: Theory, Practice, and Politics*. Malden, MA: Wiley-Blackwell.

Göçek, F. M. (ed.). (2018). *Contested Spaces in Contemporary Turkey: Environmental, Urban and Secular Politics*, 1st ed. London; New York: I. B. Tauris.

Gürses, H. and Howison, I. E. (eds). (2019). *Animals, Plants, and Landscapes: An Ecology of Turkish Literature and Film*, 1st ed. London: Routledge.

Harris, L. M. (2012). State as socionatural effect: Variable and emergent geographies of the state in southeastern Turkey. *Comparative Studies of South Asia, Africa and the Middle East*, 32(1), pp. 25–39.

İnal, O. and Köse, Y. (2019). *Seeds of Power: Explorations in Ottoman Environmental History*. Cambridgeshire: White Horse Press.

İrhan, M. C. (2016). Understanding Images of Environmental Resistances: Village Women from "Gökova" to "Yesil Yol". Unpublished MA thesis. Ankara: Bilkent University.

Lewis, C. S. (1943). *The Abolition of Man*. Oxford: Oxford University Press.

McCarthy, J. (2019). Authoritarianism, Populism, and the Environment: Comparative Experiences, Insights, and Perspectives. *Annals of the American Association of Geographers*, 109(2), pp. 310–313.

Özbay, C., Erol, M., Türem, Z. U., and Terzioğlu, A. (eds). (2016). *The Making of Neoliberal Turkey*. London; New York: Routledge.

Özkaynak, B., Aydın, C. İ., Ertör-Akyazı, P., and Ertör, I. (2015). The Gezi Park Resistance from an Environmental Justice and Social Metabolism Perspective. *Capitalism Nature Socialism*, 26(1), pp. 99–114.

Sayan, R. C. (2017). Urban/Rural Division in Environmental Justice Frameworks: Revealing Modernity-Urbanisation Nexus in Turkey's Small-Scale Hydropower Development. *Local Environment*, 22(12), pp. 1510–1525.

Simonian, H. (2007). *The Hemshin: History, Society and Identity in the Highlands of Northeast Turkey*. London; New York: Routledge.

Swyngedouw, E. (1999). Modernity and Hybridity: Nature, Regeneracionismo, and the Production of the Spanish Waterscape, 1890–1930. *Annals of the Association of American Geographers*, 89(3), pp. 443–465.

———. (2015). *Liquid Power: Contested Hydro-Modernities in Twentieth-Century Spain*. Cambridge, MA: MIT Press.

Turhan, E. (2017). Right Here, Right Now: A Call for Engaged Scholarship on Climate Justice in Turkey. *New Perspectives on Turkey*, 56, pp. 152–158.

van Dooren, T. J. (2018). Environmental Humanities. In N. Castree, M. Hulme, and J. D. Proctor, eds, *Companion to Environmental Studies*. London; New York: Routledge, pp. 418–422.

Warde, P., Robin, L., and Sörlin, S. (2018). *The Environment: A History of the Idea*. Baltimore, MD: Johns Hopkins University Press.

Yaka, Ö. (2017). A Feminist-Phenomenology of Women's Activism Against Hydropower Plants in Turkey's Eastern Black Sea Region. *Gender, Place & Culture*, 24(6), pp. 869–889.

Part one
Landscapes on the edge

2 The soils of Turkey

Nature, science, and crisis (1930–1960)

Seçil Binboğa

May 27, 1960, marked Turkey's political history in a critical way. That morning, Turkey's first military coup took place and the *Milli Birlik Komitesi* (MBK; National Unity Committee) took over the Democrat Party (DP) government. On September 16–17, 1961, the prime minister and ministers of foreign affairs and finance were executed. The military held trials for offenses such as treason, embezzlement, and violating the constitution. Among the trials was a local case of misconduct that a retired cadastral officer from the southern city of Adana brought to the MBK's attention. The case was entitled, *The Failed Attempt to Establish a Sugar Factory in Adana* (BCA 10-9-0-0 316/982/1). In his petition to the MBK, the cadastral officer claimed that this factory's failure was an example of moral bankruptcy, and he attached a brief description of how the ruling party had wasted national resources in an effort to remain in power (BCA 10-9-0-0 316/982/3: 4).

The ultimate accusation was that the DP government had spent state treasury funds to acquire largely fictitious assets with the aim of consolidating their constituency in Adana. The petition portrayed a derelict site lording over large tracts of land, expropriated to establish a factory for the manufacture of cane sugar. The cadastral officer argued that the DP government had even organized a groundbreaking ceremony on the verge of the 1957 general election and hired some administrative staff for this factory. However, there were no signs of a factory by 1960, other than a few abandoned buildings possibly constructed for housing prospective workers and staff (Ibid.).

The MBK took these accusations into consideration, and immediately ordered the Ministry of Industry's Investigation Committee (*Sanayi Bakanlığı Tahkik Heyeti*) to conduct an investigation into the ghost factory and collect forensic evidence about the surrounding landscape. The primary defendants were the former Prime Minister Adnan Menderes, the former Minister of Industry Samet Ağaoğlu, and the former General Director of the *Türkiye Şeker Fabrikaları A.Ş.* (TŞFAŞ; Turkish Sugar Factories Inc.), Baha Tekand. After pre-trial investigations and interrogations, on December 4, 1960, the Higher Investigation Commission (*Yüksek Soruşturma Kurulu*) formally charged the defendants with "burying a tremendous fortune like the State's 21 million liras in the arid soils of Adana for an investment that had proved to be unprofitable" (BCA 10-9-0-0

316/982/1: 45). Accordingly, the defendants were detained for pending trials at the *Yüksek Adalet Divanı* (YAD; Supreme Court of Justice).

What distinguished this seemingly tangential case from the larger series of scandalous allegations brought on by the military coup was the core question it evoked: "What was land?" This was neither a simple case of land-use fraud, concerning the distribution of land grants, nor was it solely an expropriation dispute limited to the realm of property rights. The question of what land is brought to the fore a complex puzzle about how to conceptualize nature at the intersection of politics, industrialization, and agriculture.

The investigations traced the case back to the 1930s, when the cultivation of sugar cane in Turkey first emerged as an experimental project, in the larger context of the state-sponsored research into the modernization of agricultural technology and practice. Central to the processes of state formation was the discovery of a new nature grounded in scientific norms of objectivity, i.e. a nature that existed insofar as it was inscribed on textual, visual, and technical records of "global environmental knowledge" (Selcer, 2018; Edwards, 2010), such as temperature and rain gauges, soil maps, and aerial photos. During the single-party period (1923–1945), the ruling Republican People's Party (RPP) inaugurated the institutionalization of environmental knowledge through numerous experiments on soil, vegetation, and irrigation. Small experimentation fields, large experimental farms, and crop breeding stations proliferated across the country (Tekeli and İlkin, 2004: 281–362). After the transition to the multi-party period, the DP government took over these experiments (1950–1960).

A systematic sugar cane experiment was initiated in 1939 by the TŞFAŞ, a state-owned enterprise whose records illuminated the following questions I ask to make sense of the 1960 investigations: How did the experiment change the meaning of nature and inform the aspirations for agricultural development and industrialization? What agents helped to facilitate such an experiment? What sorts of practices, technologies, and materials were deployed? How did each practice, technology, and material entail a peculiar mode of knowing land and sensing nature? What did the experiment results show, prove, or legitimate? And finally, how did this experimental project from the 1930s become a source of ambiguity and allegation almost three decades after its implementation?

Pursuing these questions, this chapter aims to demonstrate how various actors including scientists, politicians, and engineers contributed to the fabrication of an "experimental nature." I focus on a specific form of evidence operationalized in the court records: the scientific survey reports produced by sugar technologist Cornelis van Dillewijn (1939), soil specialist Willem van Liere (1947), agronomist Leo Placide Hebert (1952), and soil specialist Harvey Oakes (1952–1954). Through a close examination of these reports, I argue, one can discern the lineaments of an alternative history of the built environment that corresponds to a peculiar form of politics, stratified in the soils of Turkey.

I will first probe the two initial surveys to reflect on how soil became evidence for rendering land "visible," "legible," and "measurable" as a spatial object of inquiry for future experiments (Scott, 1998; Li, 2014). Second, by analyzing the two subsequent surveys I will explain why the sugar cane experiment came to a halt in the politically and financially conflictual context of the 1950s, out of which land emerged simultaneously as a "fictitious commodity" and "natural resource" (Polanyi, 1944; Cronon, 1994). My goal is to rethink these entwined conceptions of land as instantiations of an experimental nature under continuing construction.

Discovery: *saccharum spontaneum*, the Mediterranean, and the Adana Delta

This section explores the state-sponsored "discovery" efforts in the 1930s and 1940s to formulate a scientific program for the sugar cane project. Discovery, in a way that it sorted out soils, waters, and plants of a particular locale, provided the initial formulation of an experimental nature while at the same time making a "spatial history" of *the locale* (Carter, 1987). What follows is an attempt to uncover the spatial history of *the Adana Delta* written by the scientists through the tools and methods of cane research and soil mapping. I will stay with the scientists' narrative in order to understand how they collected data, performed facts, and fashioned a nature of their own (Haraway, 2003: 19). This narrative offers a penetrative inquiry into the origins of a profound change regarding the human-environment relations in the history of modern Turkey.

An origin story

The first comprehensive survey on the soils of Adana belongs to the Dutch sugar technologist, van Dillewijn, the former director of the Cirebon Research Station in colonial Java.[1] In 1939, the TŞFAŞ administration invited van Dillewijn for an inquiry into the prospects of a cane sugar industry in Turkey. The scientist compiled his notes of a three-month preliminary fieldwork into what I will refer to as *The Cane Report* (BCA 10-9-0-0 316/982/4 1939). As a reflection of the late nineteenth- and early twentieth century Dutch colonial expertise, this survey was emblematic of an "increasingly scientific (even sometimes 'scientistic') approach to the agriculture and horticulture of cane," as Roger Knight explained in the context of Java sugar's global trajectory (2000: 227).

The Cane Report argued for the possibility of establishing a sugar factory in Adana based on its three major findings and set the stage for subsequent processes of research. First of all, van Dillewijn relied on what was already visible in the field: plenty of *Saccharum Spontaneum*, i.e. wild cane. By looking at a picturesque landscape covered with three- to four-meter-long flowered stems of this tropical grass, known as one of the closely related species to *Saccharum Officinarum*, i.e. sugar cane, the scientist hypothesized that sugar cane had already been cultivated in the past or was being cultivated at the time of observation (BCA 10-9-0-0 316/982/4 1939: 6). He contended that this was a significant finding in taxonomical terms, as "Turkey" could now be added to the long list of places where wild cane had been observed, including "the island of Formosa, the Dutch Indies, and

British India" (Ibid.). This list was essentially historical rather than emblematic, as place names signified already "discovered" landscapes from which the local knowledges of cane cultivation had been extracted for colonial botany (Grove, 1995).

The second finding of *The Cane Report* was grounded in the fact that sugar cane had been cultivated in the Mediterranean region during the medieval period (see Tabak, 2008). Drawing on the historical geography of sugar production, van Dillewijn wrote: "The cane varieties, initially imported by Arabs to the Mediterranean, extended over a wide range of geography (including Spain, France, Cyprus, Anatolia, Egypt, and Morocco), making these places major centers of sugar cane agriculture, during the thirteenth and fourteenth centuries" (BCA 10-9-0-0 316/982/4 1939: 6). And indeed, sugar production in the Mediterranean at the time was "a school for the colonizers of Madeira, the Canaries, and tropical America" (Galloway, 1977: 177).

Van Dillewijn's third major finding reinforced the previous ones. Farmers were still cultivating sugar cane on the Mediterranean coasts of Anatolia, predominantly in the Adana Delta – in separate fields collectively comprising around 30 hectares (BCA 10-9-0-0 316/982/4 1939: 6). This cane was not sold commercially, but consumed by chewing and cooking (Ibid.). Van Dillewijn noted that three different varieties were being grown (Ibid.: 8). The two varieties belonged to "Zwart Cheribon," a breeding cane extensively cultivated in Java during the nineteenth century, and which the author surmised had probably been imported from Egypt (Ibid.). These varieties were distinctively soft and juicy, which made them a tasteful food. Nevertheless, they were very vulnerable to diseases and temperature changes. There was one more variety, "POJ 105," observed only in one field, which was neither suitable for chewing nor cooking due to its hard skin (Ibid.).[2] Unlike edible varieties, however, POJ 105 proved to be more tolerant of difficult climatic circumstances and produced better yields (Ibid.). Van Dillewijn traced the trajectory of POJ 105 from the Dutch East Indies to the Mediterranean back to 1902, when a Belgian technician in Java, Henri Naus, had imported some samples to Egypt (Ibid.).

After his early career in Java, Naus (1875–1938) started working at a French-owned sugar company in Egypt (Kupferschmidt, 1999: 11). The company maintained a monopoly over sugar production by implementing the "modern factory type" (Owen and Pamuk, 1998: 32). Naus, who had been hired as a factory manager, eventually accumulated considerable wealth, becoming an industrialist and resident of Egypt (Kupferschmidt, 1999: 11). This was in large part due to the great agricultural success of the POJ 105 samples that he had ordered from Java, his former area of specialization (Ibid.: 43–44).[3] Evidently, the Turkish state embraced this success story as an inspiring example for the improvement of Turkey's local sugar industry. Therefore in 1930, the Turkish Ministry of Economy (*İktisat Vekâleti*) mediated the import of some POJ 105 samples from Egypt for the purpose of experimentation in the fields of Adana Agriculture School (*Adana Ziraat Mektebi*). This was a year-long, promising experience (Alataş, 1931).

In the tracks and reports of these earlier attempts, van Dillewijn discovered a larger cause: the pursuit of a long-term research project. In other words, the scientist saw in the local state's "environmental imaginary" (Mitchell, 2011) and related efforts to engage with a global circulation of crop samples the possibility and necessity for future investment in constructions of nature.

Constructions of nature

The main chapter of *The Cane Report*, "Soil, Climate, and Irrigation" depicted, measured, and evaluated the natural and environmental forces in Adana and its environs in order to create a spatial model of the field site (BCA 10-9-0-0 316/982/4 1939: 2–8). This was conducted by juxtaposing different sets of data, derived from history of climate change, geologic maps, and attempts, if any, at irrigated farming. For instance, based on the climate data from the previous decade, van Dillewijn argued for the classification of a long strip that lies between the Taurus Mountains and the Mediterranean coastlines (roughly from Antalya to Hatay, including but not limited to Adana) within the list of sub-tropical zones convenient for cane agriculture, such as "Northern parts of British India, Egypt, Spain, Louisiana, and Argentina" (Ibid.: 11). Despite needing more longitudinal data, his calculations indicated that the growing season for sugar cane – a crop vulnerable to frost – was limited to a window between April and November in the Adana Delta.

The next major dimension required for modeling the field site was the knowledge of soil based on a soil survey. In the absence of such a survey, van Dillewijn made use of the data taken from geologic maps provided by the Institute of Geology and Mineralogy of Turkey (*Ankara Jeoloji ve Mineroloji Enstitüsü*) (Ibid.: 3). Through an analysis of this cartographic data set, the author portrayed the soils of Adana as the product of a geologically young delta system. This had been formed by alluvial sediment from the Seyhan, Ceyhan, and Berdan rivers, which for centuries drained watersheds from the Taurus Mountains into the Mediterranean. In order to make a fine-grained analysis of the soil properties and classify the soil types, van Dillewijn advised the TŞFAŞ administration to obtain a detailed analysis of sedimentary material through the expertise of a soil scientist (Ibid.). This was vital to the future planning of an irrigation scheme, one that was indispensable for cane agriculture in Adana, especially considering its extremely arid summers. It was not until the end of WWII, however, that an in-depth soil survey could be conducted.

Meanwhile, the TŞFAŞ imported new samples of POJ 105 along with a series of breeding materials from Java, Louisiana, and India. The studies on breeding continued under the management of local agricultural engineers. Van Dillewijn returned to Turkey in 1945, after a leave during WWII, and took over ongoing experiments. With him, Dutch soil scientist van Liere joined the research team for a soil survey and mapping of the field site.

Van Liere's survey was a critical example of the modern field of soil science. He compiled not only his findings but also visual documentation of survey methods and mapping processes into what I will refer to as *The Delta Report* (1947). According to him, this report was a means to translate the experience of a particular locale to make it relevant in other locations with the same or similar soil types and landforms. In terms of agricultural practice, it would then be possible to avoid risks and experimental errors, as the circulation of environmental knowledge encapsulated in soil maps would enable scientists to better predict the course of natural events (Latour, 1999: 24–80).

The designated site for the survey consisted of lands from three different villages, Denizkuyusu, Kayarlı, and Abdioğlu, located in and around a fertile area known as the Yüreğir Plain, between the rivers of Seyhan and Ceyhan. In addition to these lands' detailed soil maps, made at a scale of 1:10.000, a general agro-geological map of a larger portion was drafted based on 1:100.000 topographic maps provided by the Turkish General Command of Mapping (*Harita Genel Komutanlığı*). This larger portion corresponded to the plains bordered by the Berdan River in the West; the Ceyhan River in the East; three historic settlements, Tarsus, Adana, and Misis in the North; and the Mediterranean Sea in the South.

The agro-geological mapping aimed to historicize the category of "fertile land," as fertility was not only geological but also historical. For this purpose, the scientist treated each landform – riverbank, basin floor, dune, marsh, dike, lagoon, shore, hill or *höyük* (mound) – as a unique form of knowledge. This would in turn allow for the Delta to be read as a deeply historical terrain of superimpositions, entanglements, or segregations between nonhuman and human forms of land-making. For instance, van Liere noted that it was possible to trace settlement patterns back to the Roman period by looking at the ways in which the present villages were located on the shores of riverbanks (1947: 37). He also discovered *höyüks* and found many pieces of cereal pot in the surveying of these prehistoric sites of inhabitance. These findings illustrated that the Delta had long since been a barn of grain and other crops (Ibid.: 38). Exploring the entire field as an "archaeological landscape," rather than simply alluvial sediment, van Liere ultimately suggested that the soils of Adana were in large part composed of "anthropogenic soils," meaning, these soils were the outcome of centuries of intense agriculture and settlement practices.[4]

In order to extract information about the original natural characteristics of soil, the scientist had to conduct a series of in situ material analysis. This would provide the necessary data for the detailed soil maps. The field assistants were organized into teams of two (Figure 2.1), each team drilling one-meter-deep holes in order to track differences by analyzing vertical sections of the soil, called "profiles." The soil profile comprised the stratification of multiple layers of different qualities and types of soil, its "horizons." Soil maps were made based on the data collected from the upper horizons (one to two meters) of the earth's surface, "crumb." Due to the harsh climate of Adana, van Liere advised the teams to first hack 30 cm-deep holes with a hoe so that the dry crumb could be removed, and then continue drilling. Teams drilled one hole per hectare. If there were

Figure 2.1 Photo 3–4–5: A mapping group at work (van Liere, 1947: 6)
Source: Institute for Soil Survey, Wageningen University Collection

significant differences in the soil profiles, then the distance between the holes was decreased in order to more precisely measure the change (van Liere, 1947: 6–7).

The teams spent the night in a tent camp, which relocated every three days. In this way, they covered approximately 14,000 hectares in almost six weeks. Field assistants took notes about the differential qualities of soil, such as weight and color. A total of 700 soil samples were collected from the most typical locations and sent for examination to a soil research laboratory in Groningen and a geological research laboratory in Wageningen. In his critical assessment of the fieldwork, van Liere noted, "the execution of soil mapping with the help of assistants without special training can certainly be called an experiment. The fact that it was possible to carry out an important part of the mapping . . . should be attributed to their knowledge of agricultural practice" (Ibid.). Finally the author compiled detailed soil maps by juxtaposing the local knowledge of agricultural practice with the findings from the laboratory analysis. These maps helped to correlate the soils of Adana with the soil types in the classification system designed by the Soil Survey Institute of Wageningen. Accordingly, van Liere measured the approximate amount of land that was favorable for the cultivation of sugar cane (1947: 15–20).

Based on this elaborative survey, a model irrigation network was implemented and worked to solve the problem of aridity during the low-rainfall season. The POJ 105 samples, newly imported to tackle the risk of frost, proved to be tolerant. In January 1946, sugar technologist van Dillewijn prepared a second report,[5] comparing the results of ongoing crop experiments to the earlier assumptions he made in *The Cane Report* (1939). The latest results indicated that the advice he had given on the techniques of cultivation helped to improve the conditions of production. As a result, it became possible to produce nine tons of sugar per hectare – in contrast to four tons per hectare that he predicted in 1939 (BCA 10-9-0-0 316/983/2: 53). To conclude, the conditions for the cultivation and harvesting of sugar cane were ripe. The scientists' nature was ready to yield commodities.

All of these positive indicators encouraged more research and legitimized the formal establishment of the Adana Research Station (*Adana Şeker Etüd Bürosu*). Several scientists were recruited from the Netherlands. A chemical engineer introduced techniques for processing cane juice into crystalline sugar in order to facilitate small-scale tests of sugar production in a specially designed laboratory setting (BCA 10-9-0-0 316/982/4 1952: 28–29). Moreover, a sugar technologist reported on the possible scenarios for establishing industrialized manufacture, and explained the tightly connected relations between the plantation, the factory, and the market, through which cane sugar had emerged as a leading industrial commodity (Ibid.). Finally, an irrigation engineer designed comprehensive plans for the future management of water (De Kat, 1947). Taken together, these efforts proved that it was possible to design ideal environmental conditions for manufacturing cane sugar, albeit at the experimental scale. The next step, van Dillewijn advised, was to expand the scale of cultivation to plantations and to establish the first factory by the end of 1948 (BCA 10-9-0-0 316/983/2: 51–53).[6] And yet that next step was never taken as will be explained in the next section.

Crisis: facts, fiction, and speculation

This section explores the processes resulting in the cessation of the sugar cane experiment in the context of the U.S. Mission's Technical Assistance Program. Initially, I will focus on the final report prepared by American agronomist Hebert (1952) and briefly outline the ways in which he acknowledged, utilized, or dismissed the former narrative constructed by Dutch scientists. Moreover, I will trace how his report was co-opted by the TŞFAŞ and the Mission for their shared purpose of demonstrating the potential risks associated with the hypothetical cane sugar industry. Lastly, I will pursue the resurrection of the sugar cane narrative in a 1957 speech, delivered by Prime Minister Menderes as a campaign ploy in response to the impending economic doom. I aim to clarify how facts and fiction were mutually constitutive of the commodification of nature in the 1950s, a process during which land increasingly became an object of speculation.

Land and labor

On June 26, 1951, the TŞFAŞ administration sent a memo to the Ministry of State Enterprises (*İşletmeler Bakanlığı*) (BCA 10-9-0-0 316/982/3: 5–6). The administration was satisfied with the latest results, which proved that the Adana Delta was by and large suited to the cultivation and harvesting of sugar cane. Yet they doubted if these results would be substantial enough for undertaking new ventures in Turkey's sugar industry, one that had been specialized in manu-facturing beet sugar since the 1920s (Ibid.). In other words, the TŞFAŞ seemed reluctant to risk resources on a new product. It was also reported that the Adana Research Station had already downscaled experimental activities; but to justify a complete cessation, it was vital to make a final assessment of the twelve-year-long intensive research (Ibid.). The memo concluded with a series of questions, briefly summarized as follows: "To what extent was it sound to establish a sugar cane industry in Turkey, considering the energy needs of a factory, specificities of sugar plantations, and closest routes to the markets? At which location, if constructed, would a factory be most profitable? What sorts of machinery and equipment did the processing of sugar cane require?" (Ibid.).

Seeking answers to these questions, the TŞFAŞ approached the U.S. Mission with a request for technical assistance, asking for experts in the agriculture of sugar cane and the manufacturing of cane sugar. In response, the Mission sent the agronomist Hebert, who was then working at the U.S. Sugar Plant Field Station in Houma, Louisiana. Spending one month in southern Turkey from May 5 to June 7, 1952, and focusing on the similarities between the alluvial soils of two set-tings, the Adana Delta and southern Louisiana, Hebert prepared a comparative analysis, which I will refer to as *The Final Report* (BCA 10-9-0-0 316/982/4 1952).

The Final Report did not qualify as an in-depth fieldwork so much as a revisit-ing of earlier studies. The author was impressed by the comprehensiveness of former surveys and scientific achievements (Ibid.: 29). For instance, a viral dis-ease, "mosaic," which caused great harm to the Louisiana sugar industry in the 1920s, was completely eliminated in Adana (Ibid.: 36). And the average amount of experimental production per hectare had dramatically increased from 1939 to 1947, exceeding the amount of average commercial production in Louisiana (Ibid.: 44). However, the report identified considerable problems and downsides, especially regarding issues about the control of the hydrological cycle. Hebert noted that the Seyhan River caused extreme flood damages in winter; and unlike Louisiana, summers in the Adana Delta were hot and arid with an alarmingly low amount of rainfall (Ibid.: 45). A comprehensive infrastructure for the manage-ment of water at the delta scale (levees, ditches, and irrigation channels with a drainage system) was vital to the cultivation of cane under these circumstances (Ibid.). Yet none other than a partial irrigation network existed at the time.

In addition to the management of water, Hebert drew attention to another fundamental issue about guaranteeing access to land and labor to satisfy the daily and annual needs of a plantation economy (Ibid.: 46–47). The sugar plantation was a unique outcome of "Cheap land and labor," i.e. the exploitation of "discovered" lands and predominantly unfree labor (Moore, 2016). But, neither land was cheap, nor labor power was perpetual in the Adana Delta. Large landowners dominated the most fertile soils in the area mapped by van Liere. The leading cash crop was cotton. As Ekrem Oktar, an experienced agricultural expert of the Adana Research Station, argued, several attempts were made to convince the landowners to sell their lands to the TŞFAŞ (BCA 10-9-0-0 316/982/2: 49).[7] Yet none succeeded. Moreover, the Farmers Association-Adana Branch (*Adana Çiftçi Birliği*) encouraged the landowners to alternate cotton with sugar cane instead of selling their lands. For instance, Fazlı Meto, the head of the Association and the former mayor of Adana (1946 to 1947), noted that sugar cane could easily be integrated into the existing crop pattern, which would not only improve the local agricultural practice but also provide the national manufacturing sector with a new export product (BCA 10-9-0-0 316/982/3: 24–25). However, Hebert argued that in the absence of a market analysis on the price relations and competitiveness between cotton and cane sugar, it would be too risky to introduce an extremely labor-intensive crop like sugar cane (BCA 10-9-0-0 316/982/4 1952: 46).

Claims made by the top technocrats of the TŞFAŞ were particularly revealing in that the politically loaded term "plantation" became instrumental to legitimize their reluctance to invest in sugar cane (BCA 10-9-0-0 316/982/3). For instance, in another memo signed by the TŞFAŞ administration in February 1953, a critical statement read: "It seems doubtful whether sugar cane can settle and survive among other crops in an intensive-agriculture region like [the Adana Delta], since cane cultivation has been a matter of tropical economics and required lower standards of living in terms of agricultural labor" (Ibid.: 12). The statement continued: "We are concerned that it is difficult for this crop, which demands a *colonial labor regime* (emphasis added), to find essential economic circumstances in our country that is determined to reach higher standards of living" (Ibid.). The reasoning behind this statement had to do with the TŞFAŞ's concern about the lack of cheap and consistent supply of labor. The Adana Delta was a center of *seyyar*, *muvakkat*, or *göçebe* (mobile, temporary, migrant) labor, since the second half of the nineteenth century, when the commodification of nature based on cotton production gathered momentum (Toprak, 1997; Kasaba, 2009; Çınar, 2014). Labor was scarce other than seasonal peaks exploited by cotton farming and manufacturing (Toksöz, 2010; Gratien, 2015).

Finally, in May 1953, the chief of the U.S. Mission in Turkey, Leon Dayton, approved the cessation by stating that the Mission dismissed the sugar cane project as unprofitable especially because of the lack of labor power, and risks associated with cane sugar, a commodity highly susceptible to world market fluctuations (BCA, 10-9-0-0 316/982/3: 15–16). According to Dayton, alternating cotton with sugar cane might have been detrimental to the successful trajectory of cotton production (Ibid.).[8] As a result of these discussions that coalesced around *The Final Report*, the Adana Research Station was shut down. The cane varieties and related technological artifacts were transferred to the Ministry of Agriculture as historical relics representing the 1930s' "statist" and "protectionist" model of industrialization (Kuruç, 1988; Birtek, 1985). The fields of experimentation were sold to large landowners like the Sabancı family, who possessed extensive cotton estates in the Adana Delta, and later founded one of the largest industrial and financial conglomerates in the country (BCA 10-9-0-0 316/983/2: 9).

Reconstructions

This was not the end of the story, however. After four years of cessation, the sugar cane experiment surfaced once again, not coincidentally before the 1957 general election. The financial crisis was on the horizon. As scholars of modern Turkey have widely noted, populist economic policies, dependent upon the expansion of "cultivable land," reached their limits, resulting in high inflation, a trade deficit, and an increasing foreign debt (Karpat, 1972; Keyder, 1987; Ahmad, 2008). In the midst of a deep social unrest, the cane narrative and such could efficiently serve the Machiavellian maneuvers of an increasingly authoritarian government. On May 26, 1957, Prime Minister Adnan Menderes, in his address at the open-ing ceremony of the Adana Cement Factory, announced that the next ceremony would be held soon for the long-awaited sugar factory (BCA, 10-9-0-0 316/982/3: 28–31). This announcement triggered a puzzling set of attempts at restoring the facts retained from the past practices of discovery in an effort to fictionalize a new narrative for further investment.

The TŞFAŞ hastily prepared a cost management plan, indicating that it was impossible to conduct this project by using the funds allocated for the TŞFAŞ alone (Ibid.: 32–40). The overall implication was that the sugar cane project could be plausible only if the capital was provided by private enterprise. There-fore, a series of meetings were held with the members of the Adana Chamber of Commerce (*Adana Ticaret Odası*) and the deputies of Adana in order to create a pool of private investment to support this "new" business start-up (Ibid.: 39). Meanwhile, the TŞFAŞ acquired new lands to resume field experiments (BCA 10-9-0-0 316/983/2: 35–38). An agricultural engineer and a chemist were sent to Louisiana under the rubric of the Technical Assistance Program in order to explore the fundamentals of the sugar cane industry (Ibid.: 44). A newly formed research team undertook an expedition to determine the most convenient loca-tion for the sugar factory (Ibid.). The TŞFAŞ even invited tenders from foreign investors specialized in the technological systems and installations for establish-ing a sugar factory (BCA 10-9-0-0 316/983/1: 36–46). All of these events, which earned coverage in local newspapers, sparked rumors about the changing value of land (*Seyhan*, November 13, 1950: 1; *Seyhan*, August 19, 1957: 1). This was one of the very concrete consequences of the cane narrative: land was made into a political and cultural matter of speculation, or rather a "fictitious commodity," grounded in its legal status as alienable property.

This brief moment of speculation was interrupted by the 1958 Stabilisation Program implemented by the International Monetary Fund (IMF). This meant the devaluation of the Turkish lira from 2.80 to 9 to the U.S dollar, which made it impossible for the TŞFAŞ to account for any further progress in the sugar cane experiment (BCA 10-9-0-0 316/982/1: 24). Curiously, however, the research activities continued until the 1960 Military Coup. The YAD interpreted these activities "as originating from pure political interests, far from economic reasoning," and convicted the Democrats of "yet another corrupt act in their attempts to seek votes" (Ibid.: 45).

Tekand was jailed until October 1961, whereas Ağaoğlu had to wait for the 1963 Amnesty. Menderes had already been sentenced to hang for other convictions. The remaining defendants, such as ministers, deputies, top technocrats of the TŞFAŞ, and executives of state-owned and private banks, were found not guilty despite their contradictory claims on the question of who was accountable for the construction of a cane fiction. Each of these defendants, specializing in economics or agriculture if not business or commerce, grounded their defense in the specificities of their fields by claiming no authority whatsoever over any decisions made beyond the boundaries of their expertise. The witness statements by local agricultural engineers and field assistants, experienced in the formation of Turkey's sugar industry since the 1920s, were compelling to hear since they provided the most lengthy and detailed chronicle of events from 1930 to 1960. In spite of their immense capacity for doing deeds and recording minute details that constituted the empirical basis of the case, those engineers and assistants claimed that they had simply followed the orders imposed from above to satisfy technical means to economic ends that were never fully comprehensible to them (for an example see BCA 10-9-0-0 316/982/2: 51).

To conclude, *The Failed Attempt at Establishing a Sugar Factory in Adana* illustrated the interdependence of facts and fiction in the making and unmaking of a national economy. The whole venture started as a scientific experiment, a carry-over of a longstanding colonial "discovery" – that of the Dutch in the tropics – into 1930s' Turkey. Formulating and securing an "experimental nature" was vital to the capitalist state formation. The Adana Delta, as one example among many, was measured, mapped, and tested via models, maps, and laboratories. Experimental nature was not one monolithic construction, but many. The fact that these constructions were co-opted by various actors to invent new narratives – either factual or fictional – was not an exception to the rule. And constructions of nature persisted in the local political rhetoric, although the sugar cane experiments succumbed to the hegemonic narrative of cotton, written by the U.S. Mission in collaboration with local landowners. The distinctiveness of the 1950s was about a scalar shift in the commodification of nature, grounded in the Mission's conceptualization of land. The story would be incomplete without looking at this shift more closely.

Scale: *the soils of Turkey*

This section explores how the Mission's Technical Assistance Program took over the practices of discovery in an effort to reconstruct the category of "cultivable land," and how a nation-scale soil survey, *The Soils of Turkey* (1952–1954) could have served this purpose. This survey built on the knowledge of former studies like those conducted in the Adana Delta. Yet, the spatial scale of inquiry shifted from the regional environment of Adana to the national geography. This critical shift had to do with an emerging global concern regarding how to categorize land as a "natural resource." The technical assistance paradigm operated as a synthetic scalar practice that combined scientific, political, and historical attributes of the soil in an effort to strategically disentangle it from the earth so that it could act like a mobile entity, a "resource," imagined circulating at a worldwide scale.

From "pure science" to "applied science"

The U.S. Mission's advisory group to the Turkish Ministry of Agriculture explained the role of technical assistance as "to close the wide gap . . . between agricultural educational institutions, departmental experiment stations, and the practical farmer" (RAC R2002a/105/804, RG 1.9-RG 1.15: 33). By alluding to earlier experiments at key locations such as the Adana Research Station, the Mission reported: "Pure science has its place but applied science is the crying need just now. Coats must come off, white collars be laid aside, and ditches dug from the reservoirs of knowledge to the fields of the farm . . . until the nation, including the farmers, have attained a much higher standard of living" (Ibid.: 33–34.) These statements aimed to legitimize the need for an agricultural extension network that would bridge the knowledge of experiment stations to rural communities in an attempt to combine it with "no less valuable practical experience of the farmer" (Ibid.: 32). Conceptually, the anticipated shift from "pure science" to "applied science" referred to the *extension* of an already standardized knowledge to the locale, unlike the discovery efforts illustrated in Dutch scientists' narrative of the *extraction* of local knowledge.

The attempts to establish an agricultural extension service began in 1950, when Elmer Starch, aka "Mr. Great Plains" of the U.S. Department of Agriculture (USDA), formulated the Marshall Plan–sponsored *Technical Assistance Project No.77–26*. Initially designed as a two-year assignment, the project turned to be an exhaustive inquiry into the question of "what land is" by covering seven years of variegated field surveys (on soil, irrigation, farm machinery, cotton, livestock, grass and forage). Initially, fourteen American experts were brought in, while twenty-eight Turkish experts were sent to the U.S. for exchange of knowledge. Among the fourteen experts who came to Turkey in 1952 was Harvey Oakes, a soil surveyor from the Texas Agricultural Experiment Station. Not coincidentally, his spatial and technical focus of expertise initially brought him to the Adana Research Station. He helped to decipher van Liere's (1947) soil survey to assist the Mission in the interpretation of the sugar cane experiments. This was not his main duty however. Oakes was rather invited for a comprehensive soil survey, which would result in *The Soils of Turkey* (1952–1954), a national soil map with a scale of 1: 1.800.000.

For the completion this map, Oakes conducted nine months of fieldwork, collecting soil samples across the country. The samples were analyzed at the newly established soil laboratory in Ankara by a group of trainees under the supervision of Robert Reitemeier, another soil scientist who was sent on a mission in the scope of the same technical assistance project. Oakes then spent twelve more months working on the laboratory results to correlate the soils of Turkey with the soil types in the system of classification and nomenclature designed by the USDA (*Soils and Men*, 1938).[9] In his report accompanying the map, Oakes reasserted the mission of soil science as "to place the objects to be classified into suitable categories . . . to study and remember their characteristics and to interpret their interrelationships . . . in order that soils may be studied in the same manner as are plants or animals" (1957: 40). In this vein the soil map signified the latest phase of imperialist conception of discovery – not only of what was seen as "less civilized" through the lens of Euro-American perspective, but also of the earth in broader scale.

According to Oakes's survey, the population of Turkey in 1952 was about 22 million and the total number of animals of all kinds was about 62 million. Almost 80 percent of human population was rural or farm population, that is, "directly dependent on some form of agriculture for a livelihood," as Oakes put it (1957: 156). About 16,000,000 hectares of the total area (77,723,000 hectares) were physically suited for cultivation and 4,000,000 hectares were for irrigation (Ibid.: 4). The rest "nonarable portion" was that of "rough broken, mountainous, steeply sloping or shallow and stony areas, mostly grazed by sheep, goats, and cattle" (Ibid.: 10). Oakes adduced the numerical representation of peoples, animals, and landscapes to claim that "there was terrific pressure on the land" (Ibid.: 150). In a country like this "where nearly 80 percent of the land area was physically unsuited for cultivation," it was essential to inaugurate a detailed soil research program, one vital to the formulation of "more efficient land use with higher sustained yields" (Ibid.: 149).

No more *terra incognita*

Local soil specialists and geographers treated Oakes's survey as an exploration of an unknown land. For instance, in an article entitled, "Contours of Soil Geography in Turkey," geographer Sırrı Erinç proudly remarked, "our homeland is no more *terra incognita* in terms of spatial and temporal relations between soil types and soil-forming processes" (1965: 1). Erinç, one of the founding fathers of the field of physical geography, was primarily concerned with defining the earth's crust in geomorphological terms. When Oakes visited the Adana Research Station, Erinç was already on the site, making a reconnaissance visit in order to investigate the alluvial morphology of the Adana Delta.[10] In this context, he noted that, Oakes's survey significantly contributed to geographical modes of knowing (Ibid.: 8–9), especially at a time when the country's precious deltas were yet to be depicted with aerial photography (see also Erinç, 1952).

According to Oakes's calculation, the alluvial soils constituted the largest and most important group of "arable soils," covering approximately 3,336,330 hectares of the total area (1957: 48). These belonged to the category of "most fertile soils" among sixteen major soil groups observed in the country. Alluvial soils were found in the narrow valleys of small and large streams or basins surrounding lakes as well as in the river deltas. At the time of Oakes's survey, almost all of the alluvial deposits in the Adana Delta were used for "cropland," where cotton was the principal crop. A considerable portion was irrigated but almost 75 percent was dry-farmed. Drawing on the findings of former surveys and the updated calculations, Oakes argued: "it is safe to estimate that prevailing yields on all of these soils can be doubled under efficient irrigation and good soil management" (1957: 156). This was a call for land reclamation and river engineering in tandem with the U.S. Mission's efforts to transform the world's delta regions into a standardized commodity at the disposal of the global heavy construction industry (Sneddon, 2015). Accordingly, the *Seyhan River Multi-Purpose Dam Project* (1953–1956) engineered the largest river of the Adana Delta into one of Turkey's earliest river basin development projects through the agency of the World Bank and the Morrison-Knudsen Construction Company.

In his reflections on such radical changes in the land, agricultural engineer Adem Karaelmas, who was at the same time a qualified technocrat in the Turkish Ministry of Agriculture, argued that "the nation's greatness and prosperity was inextricably intertwined with the knowledge of soil," rendered visible by Oakes's map "accordingly with international standards of surveying and classification" (1957: 5). In this statement, it was not easy to distinguish the scientific claims about soil as a universal terrain of knowledge production from the territorial anxieties grounded in the context of ongoing nation-building processes. Beyond this ideological conflation of "terrain" and "territory" (see Elden, 2010), Karaelmas's statement was fundamentally concerned with the *resourcesness* of soil as its value became commensurable and marketable insofar as "the knowledge" pertaining to soils in one place entered a global *circulation* by means of "international" standards of map-making and classification.

The question was how to bring a geographically and territorially *fixed* material into *circulation* as an object of world-scale land speculation. An environmental history of modern Turkey can begin with this question, one to which soil science was seen as a response during the early Cold War period. Oakes's project helped to catalyze a soil research program with the help of proliferating experiment stations and agricultural extension agents working across the country to enhance techniques of soil conservation against erosion, and to develop new methods of land reclamation and irrigation. The data presented in *The Soils of Turkey* constituted the legal and institutional framework for the formation of the General Directorate of Soil Conservation and Agricultural Irrigation – abbreviated to *Toprak-Su* (Soil-Water) – within the Ministry of Agriculture in 1959. The much-debated American question of conservation – of "natural resources" – therefore became part of Turkish land policy debates. Soil's natural capabilities, rooted in its material properties and interconnected with larger cultural ecologies, acquired political significance.

Although it was designed primarily for exerting U.S. hegemony over Turkey's land-making practices, *The Soils of Turkey* also sufficiently enabled the country's transition into the international politics of development planning in the follow-ing decade. The map signified a frame of reference for a series of meetings and field surveys about the planning and management of soil, water, and forestry resources, under the auspices of the Food and Agriculture Organization (FAO) and the Central Treaty Organization (CENTO). In this context, it was first scaled into *The General Soil Map of Europe* (1:1 million) and then to *The Soil Map of the World* (1:5 million). The soils of Turkey were not inherently fixed. The fixation had to do with a series of calculations and scaled representations that mobilized the country's soils within an international process of map-making (see Craib, 2004).

Conclusion

Central to my concerns about the sugar cane experiments in Turkey was the ques-tion of "what is land." I have been occupied with the ways that western science and technology became involved in answering this question. I traced an emer-gent process of experimentation through which Dutch and American scientists together with different local actors contributed to a multilayered conceptualiza-tion of land as a scientific object, a fictitious commodity, and a natural resource. These scientists aimed to rationalize nature by classifying plant species, mapping the earth's horizons, or gauging the rivers, which served to fabricate environmen-tal narratives that have been implicated in the successive phases (and crises) of historical capitalism.

The *Cane Report* (1939) and *The Delta Report* (1947) enabled exploring a par-ticular moment in the deeply colonial legacies of "discovery," a moment in which the European imperial modes of scientific thought, through such practices like soil survey, climate analysis, and botany, intersected in formulating the underly-ing complexities of an alluvial land formation: the Adana Delta. *The Final Report* (1952) was especially important for it provoked speculative questions revolving around labor relations and commodity production, which culminated in the ces-sation of the sugar cane experiments. Finally, *The Soils of Turkey* (1952–1954) provided a medium for tracing continuity in the records of failure, crisis, and cessation – continuity in the sense that previous constructions of nature ren-dered the future visible and investible at a larger scale. The Mission's technical investment in the soils of Turkey was stratified on the basis of earlier practices of scientific discovery. In this context, the 1950s in fact encapsulated a success story, one of global capitalism, in converting land into a natural resource for a more intensive commodification.

Each survey operated as both additive and erosive by altering existing con-ceptions of land. More importantly, the ways in which each survey attended to the making of land were profoundly transformative of the understanding of *nature*, one that I considered as an expanding *experimental nature*. The concept of experimental nature allowed me to examine the Turkish state's brief flirtation with the idea of the cane sugar industry as emblematic of a joint enterprise: the long dialectic between science and capital, which has continually shaped colo-nial and neocolonial geographies of cash-crop farming. Evidence-based facts in each survey report sufficiently assimilated locality to the progressive norms of universalizing scientific practice. Nevertheless, these facts were never immune to political assertions. Various political actors with different scalar interests in land, including local state officials, representatives of the Mission, and a growing community of agrarian entrepreneurs, attempted to refashion experimental nature to their advantage.

In conclusion, the spatial outcome of historical processes of enrichment and erosion of soil was by no means identical everywhere on earth, for the locality persisted in its own peculiar forms of failure. As instantiated in *The Failed Attempt to Establish a Sugar Factory in Adana*, the seemingly banal and to a large extent redundant records of crisis could be suggestive of an extremely puzzling portrayal of a landscape, one that emerged out of factual and fictional narratives on nature.

Acknowledgements

Special thanks to William Glover, Andrew Herscher, Raymond McDaniel, Daniel Williford, Zehra Hashmi, and Dzovinar Derderian as well as Onur İnal and the two anonymous reviewers for their valuable contributions. Research for this chapter was funded by the International Dissertation Research Fellowship from the Social Science Research Council, and with research grants provided by the Rackham Graduate School of the University of Michigan.

Notes

1 Research stations that had proliferated in Dutch Java in the mid-1880s formalized a unique techno-spatial archetype, which was seen as fundamental to the scientific development of a global sugar industry (see Knight, 2014: 193).
2 "P.O.J." was an abbreviation of *Proefstation Oost-Java* (Experiment Station East Java), indicating the origin of cane varieties from this station (Ibid.).
3 According to Kupferschmidt (1999), POJ 105 replaced the local cane varieties in Egypt by the end of 1910.
4 In the 1950s and 1960s, van Liere continued his career as an expert working for the UN's FAO and conducted the surveying and mapping of the soils of Syria. As a soil scientist and geomorphologist, he not only analyzed land use and land capabilities for the development of agricultural productivity but also contributed to the formation of the scientific practice of landscape archaeology. I would like to thank Jason Ur for helping me trace van Liere's legacy (see Ur, 2010).
5 This report was attached to documents that Attilâ Tengirşenk, the former Assistant Director of the TŞFAŞ, presented to the court in his defense on October 31, 1960. It contains selected sections, not the whole report.
6 Van Dillewijn envisioned an industrial plan for the next four years to establish one factory each year until 1951.
7 Oktar noted that three different ministers of economy, Fuat Sirmen, Tahsin Bekir Balta, and Muhlis Ete visited Adana to negotiate with landowners in 1945, 1946, and 1950, respectively. All resulted in disagreement.
8 On the dramatic increase in cotton production from 1948 to 1952, see Tören, 2007: 211.
9 The USDA published *Soils and Men* in the aftermath of the Dust Bowl, and partly in response to the environmental crisis unfolding throughout the 1930s in the Great Plains region – a turning point in the American history of the colonization of land

(see Worster, 1979; Cronon, 1992). *Soils and Men*, as a quintessential example of the New Deal planners' conceptualization of nature, argued that the Dust Bowl was "an extreme result of soil misuse," and that assisting farmers about the laws of nature could prevent future environmental crises (1938: 38). This argument would not be confined to New Deal–era America. Crystallized in mapping instructions, drawing guidelines, and construction techniques, it would be widely circulated by the Mission in the following decades, in its efforts to reproduce landscapes and waterscapes across the world (see Adas, 2006; Ekbladh, 2010).

10 Erinç conducted this study with his American colleague Richard Russell, who was in Turkey at the time "to find out if the Meander River (Büyük Menderes) of western Anatolia actually meandered," as Anderson noted (1975: 380).

Bibliography

Primary sources

Adana Şeker Fabrikası yolsuzluğu ile ilgili Adnan Menderes, Samet Ağaoğlu ve Baha Tekand hakkında açılan davaya ait kararname ve mütalaaname dosyası (no.1) Başbakanlık Cumhuriyet Arşivi [BCA] (Prime Ministerial Republican Archives). Ankara, Turkey. BCA 10-9-0-0 316/982/1.
Adana Şeker Fabrikası yolsuzluğu ile ilgili davaya ait sanık ve tanık ifadeleri (no. 2) BCA 10-9-0-0 316/982/2.
Adana Şeker Fabrikası yolsuzluğu ile ilgili kararnamede atıf yapılan vesikalar dosyası. (no.3) BCA 10-9-0-0 316/982/3.
Adana Şeker Fabrikası yolsuzluğu ile ilgili rapor dosyası (no. 4) BCA 10-9-0-0 316/982/4.
Adana Şeker Fabrikası yolsuzluğu ile ilgili toplantı zabıtları dosyası. (no. 5) BCA 10-9-0-0 316/983/1.
Adana Şeker Fabrikası yolsuzluğu ile ilgili tevkif müzekkereleri ve yazışmalar dosyası (no. 6) BCA 10-9-0-0 316/983/2.
Adana Şeker Fabrikası yolsuzluğu hakkında YAD'ın 20.3.1963 tarih ve 963/98 sayılı kararı ile yazışmalar (no. 7) BCA 10-9-0-0 316/983/3.
Adana Şeker Fabrikası Çalışmaları Devam Ediyor. (1957). *Seyhan.*
"Agriculture and Its Basic Relations to National Objectives" In *Project 77–26, Agricultural Advisory Group (Starch Group)* (1953). Box R2002a, Series 105, Subseries 804, RG 1.9-RG 1.15 (A83–89), Projects, FA479, Rockefeller Foundation Records (RF), Rockefeller Archive Center (RAC).
Alataş, H. (1931). *Şekerkamışı Ziraati ve Tecrübe Raporları.* Istanbul: Hilal Matbaası.
The Cane Report: Dillewijn C. V. (1939). *Türkiye Şeker Kamışı Ziraati Imkanlarına Dair Bir Rapor* (A Report on the Possibilities of the Agriculture of Sugar Cane in Turkey). BCA 10-9-0-0 316/982/4.
Çukurova'da Şeker Fabrikası. (1950). *Seyhan.*
De Kat, J. O. (1947). *Adana Sulama İşleri.* The Middle East Technical University Library (METU) Kasım Gülek Collection.
The Delta Report: Liere, W. V. (1947). *Bericht der Bodenuntersuchung und der Bodenkartierung des Adana-Deltas, Türkei.* The METU Library, Kasım Gülek Collection.
The Final Report: Hebert, L. P. (1952). *Türkiye Şeker Kamışı Tecrübelerinin Değerlendirilmesi Hakkında Nihai Rapor* (The Final Report on the Evaluation of Sugar Cane Experiments in Turkey). BCA 10-9-0-0 316/982/4.
Karaelmas, A. (1957). *Türkiye Umumi Toprak Haritasının Hazırlanması, Faydaları ve Toprak Gruplarının İzahı.* Istanbul: Işıl Matbaası.
Oakes, H. (1957). *The Soils of Turkey.* Ankara: Doğus Matbaası.

Soils and Men (1938). United States Department of Agriculture (USDA) Yearbook 1938, Washington, DC.

Ziraat Vekâleti Toprak Muhafaza ve Zirai Sulama İşleri Umum Müdürlüğü Teşkilât ve Vazifeleri hakkında Kanun. No. 7457, November 9, 1959. Resmi Gazete. [online] Available at: www.resmigazete.gov.tr/arsiv/10444.pdf [accessed on January 7, 2018].

Secondary Sources

Adas, M. (2006). *Dominance by Design: Technological Imperatives and America's Civilizing Mission.* Cambridge, MA: Belknap Press of Harvard University Press.

Ahmad, F. (2008). Politics and Political Parties in Republican Turkey. In R. Kasaba, ed., *The Cambridge History of Turkey*, 1st ed. Cambridge: Cambridge University Press, pp. 226–266.

Anderson, C. H. (1975). *Richard Joel Russell, 1895–1971: A Biographical Memoir.* Washington, DC: National Academy of Sciences.

Birtek, F. (1985). The Rise and Fall of Etatism in Turkey, 1932–1950: The Uncertain Road in the Restructuring of a Semiperipheral Economy. *Review*, 8(3), pp. 407–438.

Carter, P. (1987). *The Road to Botany Bay: An Essay in Spatial History.* London: Faber and Faber.

Çınar, S. (2014). *Öteki "Proletarya": De-proletarizasyon ve Mevsimlik Tarım İşçileri.* Ankara: Notabene.

Craib, R. B. (2004). *Cartographic Mexico: A History of State Fixations and Fugitive Landscapes.* Durham, NC: Duke University Press.

Cronon, W. (1992). A Place for Stories: Nature, History, and Narrative. *The Journal of American History*, 78(4), pp. 1347–1376.

———. (1994). Landscapes of Abundance and Scarcity. In Clyde A. Milner et al., eds, *The Oxford History of the American West*, 1st ed. New York: Oxford University Press, pp. 603–637.

Edwards, P. N. (2010). *A Vast Machine: Computer Models, Climate Data, and the Politics of Global Warming.* Cambridge, MA: MIT Press.

Ekbladh, D. (2010). *The Great American Mission: Modernization and the Construction of an American World Order.* Princeton: Princeton University Press.

Elden, S. (2010). Land, Terrain, Territory. *Progress in Human Geography*, 34(6), pp. 799–817.

Erinç, S. (1952). Çukurova'nın alüvyal morfolojisi hakkında. *İstanbul Üniversitesi Coğrafya Enstitüsü Dergisi*, 3–4, pp. 147–159.

———. (1965). Turkiye'de Toprak Çalışmaları ve Türkiye Toprak Coğrafyasının Ana Çizgileri. *İstanbul Üniversitesi Coğrafya Enstitüsü Dergisi*, 8(15), pp. 1–39.

Galloway, J. H. (1977). The Mediterranean Sugar Industry. *Geographical Review*, 67(2), pp. 177–194.

Gratien, C. (2015). The Mountains Are Ours: Ecology and Settlement in Late Ottoman and Early Republican Cilicia, 1856–1956. Ph.D. Washington, DC: Georgetown University.

Grove, R. (1995). *Green Imperialism: Colonial Expansion, Tropical Island Edens, and the Origins of Environmentalism, 1600–1860.* Cambridge: Cambridge University Press.

Haraway, D. (2003). *The Companion Species Manifesto: Dogs, People, and Significant Otherness.* Chicago: Prickly Paradigm Press.

Karpat, K. (1972). Political Developments in Turkey, 1950–70. *Middle Eastern Studies*, 8(3), pp. 349–375.

Kasaba, R. (2009). *A Moveable Empire: Ottoman Nomads, Migrants, and Refugees.* Seattle: University of Washington Press.

Keyder, Ç. 1987. *State and Class in Turkey: A Study in Capitalist Development*. London: Verso.

Knight, R. (2000). The Sugar Industry of Colonial Java and Its Global Trajectory. *South East Asia Research*, 8(3), pp. 213–238.

———. (2014). *Sugar, Steam and Steel, the Industrial Project in Colonial Java, 1830–1885*. Adelaide, SA: University of Adelaide Press.

Kupferschmidt, U. M. (1999). *Henri Naus Bey: Retrieving the Biography of a Belgian Industrialist in Egypt*. Brussels: Academie royale des sciences d'outre-mer.

Kuruç, B. (1988). *Belgelerle Türkiye İktisat Politikası, Cilt: 1 (1929–1932)*. Ankara: A.Ü. Siyasal Bilgiler Fakültesi.

Latour, B. (1999). *Pandora's Hope: Essays on the Reality of Science Studies*. Cambridge, MA: Harvard University Press.

Li, T. M. (2014). What Is Land? Assembling a Resource for Global Investment. *Transactions of the Institute of British Geographers*, 39(4), pp. 589–602.

Mitchell, T. (2011). Are Environmental Imaginaries Culturally Constructed. In D. K. Davis and E. Burke III, eds, *Environmental Imaginaries of the Middle East and North Africa*, 1st ed. Athens, OH: Ohio University Press, pp. 265–275.

Moore, J. W. (2016). The Rise of Cheap Nature. In J. W. Moore, ed., *Anthropocene or Capitalocene? Nature, History, and the Crisis of Capitalism*, 1st ed. Oakland, CA: PM Press, pp. 78–116.

Owen, R. and Pamuk, Ş. (1998). *A History of Middle East Economies in the Twentieth Century*. London: I. B. Tauris.

Polanyi, K. (1944). *The Great Transformation: The Political and Economic Origins of Our Time*. Boston: Beacon Press.

Scott, J. (1998). *Seeing Like a State: How Certain Schemes to Improve the Human Condition Have Failed*. New Haven, CT: Yale University Press.

Selcer, P. (2018). *The Postwar Origins of The Global Environment: How the United Nations Built Spaceship Earth*. New York: Columbia University Press.

Sneddon, C. (2015). *Concrete Revolution: Large Dams, Cold War Geopolitics, and the US Bureau of Reclamation*. Chicago: The University of Chicago Press.

Tabak, F. (2008). *The Waning of the Mediterranean, 1550–1870: A Geohistorical Approach*. Baltimore, MD: Johns Hopkins University Press.

Tekeli, İ. and İlkin, S. (2004). *Cumhuriyetin Harcı, İkinci Kitap: Köktenci Modernitenin Ekonomik Politikasının Gelişimi*. Istanbul: Bilgi Üniversitesi Yayınları.

Toksöz, M. (2010). *Nomads, Migrants and Cotton in the Eastern Mediterranean: The Making of the Adana-Mersin Region, 1850–1908*. Leiden: Brill.

Toprak, Z. (1997). Cumhuriyetin ilk Yıllarında Adana'da Amele Buhranı ve Amele Talimatnamesi. *Toplumsal Tarih*, 41, pp. 7–10.

Tören, T. (2007). *Yeniden Yapılanan Dünya Ekonomisinde Marshall Planı ve Türkiye Uygulaması*. Istanbul: Sosyal Araştırmalar Vakfı.

Ur, J. A. (2010). *Urbanism and Cultural Landscapes in Northeastern Syria: The Tell Hamoukar Survey, 1999–2001*. Chicago: Oriental Institute of the University of Chicago.

Worster, D. (1979). *Dust Bowl: The Southern Plains in the 1930s*. New York: Oxford University Press.

3 A technopolitical frontier

The Keban Dam project and southeastern Anatolia

Dale J. Stahl

In 1975, the Turkish government announced completion of a great dam on the Euphrates River near the town of Keban in the Elazığ Province. The massive dam was one of the largest civil engineering projects in the world in the late 1960s and 1970s, costing nearly 9.2 billion Turkish lira. Rising 163 meters above the riverbed, the Keban Dam produced a reservoir 675 square kilometers in area with a volume of 30,600 square hectometers (Akarun, 1999: 1–2; Orhon et al., 1991: 382). The reservoir became Turkey's third-largest lake, a status lost with the completion in 1992 of an even larger dam and reservoir, the Atatürk, downstream. The growing reservoir also displaced as many as 40,000 people, drowning 94 inhabited villages and an unknown number of archaeological sites (Öktem, 2002: 315). Extant analyses of the Keban Dam focus on its material presence and hydrological influence, but the history of the dam project involves much more than a story of environmental engineering (Kolars and Mitchell, 1991; see also Biswas, 1994; Rogers and Lydon, 1996). This chapter traces the genesis of the Keban Dam in the 1950s and 1960s through a series of interlocking discourses, examining not only the statements and actions of the engineers and politicians who designed, built, and funded the project, but also the writers, poets, and journalists who generated a new literary discourse to support environmental transformation. Within these discourses, the dam project functioned at and as a technopolitical frontier: a social, political, technical, and ecological threshold. Once formed, transgressed, and reconstituted, this technopolitical frontier made possible a new conception of southeastern Anatolia.

The notion of "frontier" operates in this chapter in two ways, relying first on the idea of eastern Anatolia as a frontier region, and, second, on the concept of the Keban Dam project as an inception point or boundary, as the necessary project for further imagining and engineering. The idea of eastern Anatolia as a frontier has a history extending at least to the middle of the nineteenth century and connected, at least in part, to the fact that communities in this area had over several decades resisted central government control.[1] In the late nineteenth century, Armenian resistance to the Ottoman state intensified, eventually culminating in the massive dislocation and genocide of the First World War (Suny et al., 2011; Akçam, 2013). Then, after the war, Kurdish communities in the southeast rose up against the new government of the Turkish Republic during the Shaykh

Said rebellion (Olson, 1991: 91–127; McDowall, 1997: 184–213). In addition, as ideas of Turkish nationalism developed within the Young Turk movement in the early twentieth century, eastern Anatolia came to represent a "backward" region filled with peoples to be removed, tamed, or assimilated.[2]

Similar to other territories viewed this way, eastern Anatolia came to be seen as a "backward" or "wild" region, in need of technologically driven sociopolitical change. Indeed, one may view such technical "solutions" as relying on the "frontier" designation for their invention and existence (see Davis, 2007; Davis and Burke III, 2011; Pritchard, 2012: 593–596). After the Second World War, government committees justified the Keban project as addressing longstanding economic inequities between eastern and western Turkey. Official policies viewed Kurds as proto-Turks, subject to a process of "Turkification," which would " 'push' people forward . . . towards the identity of the future" (Üngör, 2011: 139). That future, according to the Turkish government, involved a particular understanding of modernity that contemporary social theory defines today as "technopolitical," or as the "strategic practice of designing or using technology to constitute, embody, or enact political goals" (Hecht, 1998: 15).

The implementation of technological solutions, however, very often emerges from a division of the world into "neatly separate realms of reason and the real world, ideas and their objects, the human and the nonhuman," such that the "mixed way things happen" is reduced to mere binary opposition (Mitchell, 2002: 52). The "frontier" operated as such a binary, with "wild(er)ness" on one side and a bright, modern future on the other. The perceived disparities between western and eastern Turkey appeared in official and unofficial literatures in myriad ways; the 1963 Five-Year Development Plan, for example, sought "the establishment of a balanced social order" (State Planning Organization, 1963: iii). The task, then, of dam proponents was to situate the Keban project as the essential remedy to "imbalance." In the discussion that follows, we will see how the historical framework of the "frontier" appears, time and again, in a range of discourses, to facilitate the technopolitical "fix" of the Keban Dam.

Even as the "frontier" designation helped to invent a technopolitical "solution," the Keban Dam itself functioned as a threshold for the application of further technopolitical interventions. As will be discussed in greater detail below, several Turkish engineers and politicians viewed the Keban Dam as creating something of a "path dependency," such that the building of the dam and its relation to the economic conditions of southeastern Anatolia would make it both feasible and necessary to construct additional dams. It was through the Keban project, then, that a set of factors – some cultural and political, some material and geographical – were combined and arranged such as to make further environmental engineering and social reshaping possible. Put another way, the literal and proverbial "mass" of the dam supported ever more grandiose notions of southeastern Anatolia's "development." In fact, I argue that this mass enfolded the "frontier" idea in such a way to make it reusable for later development plans. So, even as the "frontier" justified the intervention of the Keban Dam, the dam and the meanings associated with it cemented the frontier model for future use.

In this way, the Keban Dam made it possible to render more and more of southeastern Anatolia within a technopolitical frame.

Dams are usually more than just a pile of rocks and cement, of course, but some dams do more than plug a river; they are conceptually productive. Recent scholarship on dam building projects in the Middle East has acknowledged this feature (e.g., Derr, 2011; Reynolds, 2012; 2017). Timothy Mitchell writes of dams as a means to "rearrange the natural and social environment . . . to demonstrate the strength of the modern state as a techno-economic power." To Mitchell, dams require "a significant reorganization and concentration of accounting, calculation, description, and knowledge" (36). This observation of dam building's sociopolitical function has given rise to historical claims regarding the proliferation of dam building technologies and ideologies. Christopher Sneddon (2015) has termed this proliferation a "concrete revolution" wherein the construction of massive dams became a vital feature of twentieth-century economic development. American expertise and technical assistance programs played a major role in bringing about the "concrete revolution." This chapter complements parts of Sneddon's story by relating the career of Turkish engineer and Prime Minister Süleyman Demirel, who worked with the US Bureau of Reclamation.

Still, the challenge of historical scholarship on dams is to consider how these models and ideas relate to particular conditions and contingencies. Even as the Keban project owed a debt to earlier, promethean undertakings – the Hoover, TVA, and Aswan – the natural, cultural, and sociopolitical environments the dam sought to rearrange contained unique features in their own right. Historians, then, cannot ignore some questions: what kinds of knowledge? Whose calculations? And so on. Each great dam project delivered unto societies powers akin to the gods, but societies produced and applied power in particular contexts with specific effects. This chapter works through three intertwined histories that reveal some of the Keban Dam's particularities, working on three interrelated scales: local, national, and international. The first section examines the cultural production of a "new Euphrates River," undertaking a discourse analysis of stories and poetry produced near the advent of dam construction. The second section analyzes biography, detailing the careers of the "engineer-politicians" who came to rule the Turkish Republic after the Second World War. The third and final section details the social life of an artifact, a feasibility report, designed to frame Turkish economy and society for the application of international development aid.

Each section of the chapter details the ideas, influences, and changes wrought by actors crucial to bringing about the Keban Dam project. A caution, then, is necessary. The "visions" and "imaginings" discussed herein must be viewed from a critical vantage, as representing the ideas of actors within or affiliated with the government, or nationalist boosters interested in seeing the dam built. One of the effects brought about by technopolitical framing is a diminishing or disappearing of stories of reaction and resistance. Indeed, direct archival material on local dissent to the Keban Dam project is scant. Still, some traces of resistance may be seen in this chapter. Efforts to change cultural narratives of the river imply an earlier vernacular that could challenge or thwart the supposed social purpose of the dam.[3] Rock and river also manifested resistance to the designs of engineers, eroding canyon walls and revealing design flaws. Though the Keban Dam stands today seemingly as testament to the failure of resistance, the connected histories of human, rock, and river continue as part of contemporary protest against other nearby dam projects (e.g., Hommes et al., 2016).

Ultimately, the Keban Dam project generated new ways of thinking and framing such that a much larger engineering and economic development program became imaginable, that is, able to be conceived and constructed. Out of the Keban Dam came the Southeastern Anatolia Project, often referred to by its Turkish acronym, GAP (*Güneydoğu Anadolu Projesi*). GAP is one of the largest multisector development schemes in the world, encompassing nine provinces of the Turkish Republic, costing more than 12 billion USD, and involving the construction of twenty-two dams (Kolars and Mitchell, 1991: 25–29). How did one dam become twenty-two? A contingent set of factors, calculations, descriptions, and knowledge, operating at multiple scales, helped to make the Keban Dam into the catalyst project for a grand development scheme. The purpose of this chapter is to detail the histories and the knowledge that were gathered, layered, and positioned to bring about the Keban Dam and its many descendants.

The poetic engineering of a new Euphrates River

The town of Keban sits near the confluence of two rivers, the Karasu and the Murat, that join to form the Euphrates River. As any glance at a topographical map will make clear, the interaction of rivers and mountains made this particular spot. The Karasu and Murat have cut their own deep canyons through the mountains of eastern Anatolia, flowing nearly one thousand kilometers before entering a relatively open and level area of land characterized by their confluence. Below this confluence, the combined river, now known as the Euphrates, has made its own long canyon which slices through the Toros Mountains. The mountainous terrain eventually gives way to the Syrian steppe; the Euphrates River leaves Turkish territory soon thereafter. The river meets no significant tributaries during its journey through Turkey to Syria.

Geography made Keban a good spot for a dam. The great canyon walls rearing up after the confluence seemed perfect for buttressing a massive construction, and the dam, if built large enough at that point, could command the floods that raged seasonally through the canyon below. Geology, on the other hand, made it a rather difficult place for a dam. Three faults striated the gorge where engineers planned to lay the north gravity dam and the powerhouse, necessitating the redesign and relocation of the powerhouse during the construction process. Nature had also seen fit to fill the gorge walls with karstic conditions, which threatened the integrity of the dam and required filling. Engineers did not find some of these karst formations until after the dam was finished and the reservoir slowly filling. For several years after the dam's construction, engineers had to fill the supposedly solid canyon walls with grout (Akarun, 1999).

While topographical formations (if not geological ones) made Keban a suit-able place to position a dam, human ingenuity was required to bring the dam into existence. As a result, cultural formations were an equally important aspect of the dam's construction. From the early 1960s, local periodicals began to fill with reports, stories, and poetry about the river and the dam. Some of this writing was overtly propagandistic, extolling the great project, the great men who would bring it about, and the great future the dam would make possible. But other works resonate more as reconciliation, as an attempt to make sense of incompatible conceptions of the environment and the human place within it.

In 1962, the arts and culture journal *Yeni Fırat* ("New Euphrates") began pub-lication in Elazığ. The purpose of the journal was "to benefit from written work in order to learn about history, geography and folklore." The editors stated that the title, *Yeni Fırat*, had been chosen for two express reasons. First, they wished to acknowledge the journal's debt to another publication, called simply *Fırat*, which began its short-lived publication history in 1918. Second, the journal meant "to describe the effulgence that remains today of the metaphor of the Euphrates, the source of life for our environment and perhaps for all of Turkey" (*Yeni Fırat*, 1962: 3). According to the journal, the river carried a range of cultural meanings along with providing a critical resource for life in the mountains and arid plains of eastern Anatolia. Another, third reason for the journal was not expressed in so many words: *Yeni Fırat* was also in some sense about making a new Euphra-tes. One might imagine, then, the journal's poets and storytellers as engineers of another kind, making not machines but artful structures of language intended to shift cultural meanings.

This new Euphrates River came into existence over the course of the 1960s in the pages of local publications. The river transformed from a symbol of nature's beauty and bounty to a destructive phenomenon that required human mastery and guidance. In the local newspaper, the *Elazığ Gazetesi*, in the late 1950s and early 1960s, several poems describe the river as part of the region's identity, a nat-ural phenomenon that brought life and prosperity. The paper printed the poem, "Euphrates" (*Fırat*), by Arif Nihat Asya. Asya (1963) describes a river that flowed from "blue mountains far away," giving its name to a beloved folk song. The river "flows into the flavor of the song (*türkü*)." Asya's narrator goes on to express his life's hope, crying out at the end, "Ve senden doğacak kızımın/Adı 'Fırat' olsun!" ("And my daughter who will be born from you/Let her name be 'Euphrates!'"). The same image, an evocation of the life-giving quality of the river, is present in the first issue of *Yeni Fırat* (1962), such that the river's name befits a newborn child.

Later poems, from roughly 1963 forward, published in the newspaper and *Yeni Fırat* describe a very different river. Yıldırım Gençosmanoğlu (1963) wrote the poem, "Reckoning with the Euphrates" (*Fırat'la hesaplaşma*), in which he took the river to task. "You split my valleys, you swallowed my plains," the narrator accuses, "You scattered the ashes of many hearths; you gave them to the wind!" The river is now a destructive and unfriendly force, "You raged and frothed, you gave my country to salt/Neither a road for my caravans nor a ford for travelers/

We depended on you; you gave us suffering." The Euphrates is, in this rendering, a harmful and resentful presence, devastating human constructions and thwarting human designs. The poem's title suggests something of what Gençosmanoğlu believes should happen to the recalcitrant river. The root word, *hesap*, suggests accounting, calculating, and planning – the very processes Timothy Mitchell notes as crucial to dam building.

Gençosmanoğlu's poem continues with imagery that connects the construction of a modern nation to the taming of the river. "Oh lion who roars down from the Palandöken [mountains]/We will put a chain around your mane/And forget the past; and you, me, shoulder to shoulder,/We will found a nation that rises with prosperity." The personal quality of Arif Nihat Asya's Euphrates has given way to collective demands in Gençosmanoğlu's version. What the river was in the past – the narrator likens the Euphrates to an oppressive sultan – can be forgotten. The river could no longer act like an Ottoman despot in this new, democratic Turkish Republic. Such a sentiment, directed not at rivers but at other features of Ottoman life, was long-standing both within Turkey and without (Zürcher, 1992; Citino, 2008).

Cenani Dökmeci was less explicit about controlling the river in his poem, "Dialogue with the Euphrates" (*Fırat'la söyleşme*), published in 1966. Dökmeci's version of control is less vehement than Gençosmanoğlu, but he reached even farther back into the history of the Euphrates River to make his claims. Rather than a reckoning, Dökmeci's "Dialogue" beseeches the river, "Stop Euphrates! . . . Don't flow and burst forth into the deserts, stop, no more." The river should instead become "an inland sea in the nation's breast." As a reservoir rather than a river, the Euphrates will be more productive. "Urfa Elaziz is very thirsty," the narrator noted, "let your shore, your contour be painted in greens" (7). Rather than watering empty wastes, a dammed river could bring relief to the pilgrims of Urfa and promote new agriculture.[4]

The dam site was mentioned specifically in the poem. "Tell me of Keban," Dökmeci wrote, asking the river to slow down and learn of the human communities along its banks. The poet presented the dam as "a new legend" (*yeni bir destan*) that may be told by the ancient river, a legend as important to Turks as Seljuq (*Selçuk*) sultan Alp Arslan's defeat of the Byzantine Empire at the Battle of Manzikert (*Malazgirt*) in 1071. Just as Alp Arslan's conquests heralded an Anatolia dominated by Turkish tribes, the dam at Keban proclaimed the construction of "an enlightened nation," capable of dominating Anatolia in an entirely new way (7). Through such writings, a new cultural imagination of the Euphrates River emerged, one that required the building of a dam at Keban. Moreover, taming the river was not simply a project for local benefit in this poetry; the great work of environmental engineering would bring about a rejuvenated nation.

Poetry was not the only method for imagining a new Euphrates. Writers also told stories to shape new, novel conceptions of the environment around Keban. One story by Bahattin Senemoğlu (1963) tells of an American who arrived in Elazığ shortly after the foundation of the Republic. In Senemoğlu's account, one day the American "went down to the shores of the Euphrates." After wandering a bit, he asks a villager about the river, "Does it always flow like this?" The villager confirms the American's observation but thinks nothing of it. Senemoğlu then claims that "the American had discovered a great treasure imprisoned in the river," a treasure that would not be recognized by Turks until much later. Only when "Turkey entered the age of planned development" did the people of Elazığ recognize what lay hidden. With the coming of the Keban project, "the eyes of the entire Turkish nation are turned toward . . . the promise of the shining sun that will rise with the construction of the dam." Though an American presence of this sort did not become widespread in the Middle East until after the Second World War, Senemoğlu projects American technical influence into the interwar period and rewards American expertise with discovering the river's potential. It is also striking that there is nothing of the tragedy of American presence as one finds in 'Abd al-Rahman Munif's *Cities of Salt*, for instance, published two decades later in 1984, nor is there any of the bilateral contestation that frequently marked relations between American and Turkish modernizers (Adalet, 2018: 85–120).

Instead, these works of poetry and storytelling constructed a new cultural framework around the river and dam that foregrounded nation-building, technical assistance and prowess, and a range of beneficial social outcomes. The river, powerful and wild, could only be tamed by the dam, a project worthy of the nation. With the building of the dam, the river's strength, wasted and destructive in flood, might be conferred on the nation itself. That these writings sought to reshape cultural meanings to justify building the dam is fairly clear. But we may see in these efforts something more than simply a public relations effort. The Euphrates was not just *any* river. It was the river that had watered some of the first human agriculture and fed the canals of great ancient civilizations. The Bible mentions the Euphrates as emanating from the Garden of Eden. Changing the river's cultural meaning meant generating a new cultural context for imagining and understanding the human engineering of the environment around Keban.

That cultural context relied on representations of the river rather different from those found in hydrological treatises; these representations very often complicated the binaries of human and nonhuman or human and nature embedded within technical inquiry. In this new framing of the Euphrates, the river became anthropomorphic and zoomorphic, represented as both human and animal. The above writers, all male, used gender and ethnicity to portray the river's anthropomorphic traits, effectively linking the river to the Turkish nationalist project. Though forms of scientific calculation may have been more directly responsible for producing the dam itself, the rhythms of poetry and the cleverness of storytelling helped produce the dam as a *cultural* object worthy of construction.

As will become clearer in the next sections, the dam became an element dependent on multiple, interlocking contexts. Some of these were nonhuman – the context of interacting water, silt, and rock – while others were human, such as the cultural context that made dam building not just an option but an imperative. Still other contexts must be layered on these, however, before a monolith of rock and concrete could rise to block the river. Poetry and local storytelling, though evocative and fascinating, were not sufficient forms of knowledge to design, build, and finance the dam. Poets and raconteurs alone could not consolidate and direct a technical bureaucracy and national government or convince international organizations to grant Turkey loans and aid for the building of the dam. At the national level, then, proponents of building the Keban Dam harnessed another kind of knowledge production. Capitalizing on a longer trend in Turkish history, these actors prioritized the language of science and engineering and connected it to a language of governance (Hanioğlu, 2005, 2017; Burçak, 2008). These patterns of thinking and discourse eventually came to occupy the very heights of Turkey's national government.

The king of dams and his viziers

Biography provides one way to analyze how science and engineering connected to bureaucracy and government in the Turkish Republic in the period following the Second World War. At this time, technical colleges became the fertile soil from which sprung much of Turkey's civilian political leadership. Sociologist Nilüfer Göle has noted the shift from a "bureaucratic elite to a technical one" in the governing of Turkey (1993: 200). One individual epitomized this trend, a man who became known in the early part of his career as the "King of Dams" (Arat, 2002: 98). Süleyman Demirel was born in İslamköy in the province of Isparta on November 1, 1924, and grew up in a political family. His father, Hacı Yahya, was later elected headman (*muhtar*) of the local municipality after Turkey's war of independence. Hacı Yahya's political career had a strong influence on his son's understanding of politics and patronage, knowledge that would serve him well later in life (Komşuoğlu, 2008: 98–100).

The young Demirel did not follow his father directly into politics, however. His upbringing in a small village encouraged other ambitions. Demirel attributed an interest in engineering to working on the family farm and growing up in a community experiencing a drought. Meanwhile, the nearby provincial center, Isparta, benefited from electrification. The young Demirel watched and was inspired by the lights flickering on in that city each night while his own village remained lit only by the stars. As a result, instead of pursuing a career in the military or working with his father, Demirel traveled to Istanbul to study hydraulic engineering at Istanbul Technical University (Arat, 2002: 87–88). While there, Demirel made contacts with a number of young men who later went on to play significant roles in Turkish politics, including Necmettin Erbakan, Turgut Özal, and his brother Korkut. All four men studied engineering, and, when they eventually turned toward politics, each joined or founded right-wing parties. This trend to the right interests Göle, who notes Demirel as a key figure in the "political ascendancy of engineers" (1993: 199) but tends to emphasize the 1980s as the period when "technocratic consciousness" met a "revived Islamist movement" (Ibid.: 206–207). Focusing on Demirel's career suggests a longer arc toward "political ascendancy," as well as the importance of other influences in shaping Turkish engineers' ideas, namely US private engineering firms and the US government.

After university, Demirel began working at the Electric Power Resources Survey and Development Administration (*Elektrik İşleri Etüt İdaresi Genel Müdürlüğü;* EİE) in May 1949. As part of his work there, he escorted delegations of American engineers sent to study and assess Turkey's hydroelectric potential. His ingenuity in ferrying US government personnel across the country brought Demirel to the attention of his superiors. In 1950, he became the first engineer sent by the Turkish government to study in the United States. For nine months, Demirel worked with the Bureau of Reclamation on water and energy projects. This visit had a significant impact on Demirel, who recalled spending three days gazing at the Hoover Dam (H. Turgut, 2000: 58). American influence was not limited only to technical knowledge. Demirel later recalled his time in the United States as fueling a new political ambition, "Water engineering's peak had been attained in the USA's western states. It was my primary goal to find an opportunity to realize in my country some of the things that I learned and witnessed in those territories" (as quoted in H. Turgut, 1992: 109).

Upon Demirel's return to Turkey, he was posted to the Seyhan Dam near Adana, a project planned by Turkish engineers in coordination with the American firm Morrison Knudsen. Recalled to Ankara in 1954 to work as a project engineer, Demirel soon became the head of the Dams Administration (*Barajlar Dairesi*) in the newly created State Hydraulic Works (*Devlet Su İşleri;* DSİ). Turkish officials deliberately modeled the new DSİ on the US Bureau of Reclamation. Demirel, one of the young engineers who underwent training at the Bureau's Denver, Colorado, office, was a natural choice for a leadership position. In this position, Demirel helped manage the rapid expansion of Turkish dam-building in the 1950s, overseeing several dams and hydroelectric projects.[5]

In 1953, a group of American businessmen established the Eisenhower Exchange Fellowships; Süleyman Demirel was a member of the inaugural group. For much of 1954, Demirel and his wife toured the United States, investigating how various US organizations operated. They cut short their trip, however, and when asked about the abbreviated visit, Demirel replied, "It was a 12-month trip. But I returned to Turkey in the tenth month. During my visits, I saw how natural resources were used, how investments that were made in this area were directed, and how money was allocated. Truly, I saw a lot of things out there. My enthusiasm, after all, was civilization [*medeniyet*]. Civilization, well; I sought it, I sought its elements" (121). An interest in technical and financial measures as "civilizing" agents placed Demirel in a long line of Ottoman and Turkish reformers, many of whom had also sought to apply such techniques, particularly to areas considered part of the "frontier" (Makdisi, 2002; Deringil, 2003; Low, 2015; Gratien, 2017).

Moreover, learning how to fashion connections between "nature, money, and modernity" was becoming at this time an important component of statecraft (Coronil, 1997). Demirel's two trips to the US showed him that the engineering sciences were not enough on their own. To produce the kinds of economic and political organization required to build "civilization" (qua dams and other waterworks), engineers needed to connect technical knowledge to other contexts and forms of knowledge. The United States, postwar superpower and economic miracle, acted as an exemplar of combining technical know-how with financial instruments and political power. After his visits to the US, Demirel saw beyond the technical processes involved in measuring the flow of water and electricity to the allocation of other flows, namely finance and capital investment (Mitchell, 2011, 2014).

One may view Demirel and other Turkish bureaucrats at this time as representative of a "managerial elite" reaching its fullness in Turkish politics. Turkey in the post-WWII period produced a whole stratum of technocrats, many in the American mold, exercising power through an ever-expanding postwar bureaucracy. But such a conception of Demirel's sociological position focuses more on social implications than environmental ones; it follows his political shrewdness more than the situated-ness of his technical expertise. Demirel meant for civilization to be literally engineered through vast, unmatched, and practically irrevocable changes in the environment. What Demirel learned in the United States was how to bend the Turkish government and the organs of international finance to that end.

Shortly after Süleyman Demirel's return to Turkey from his Eisenhower Fellowship, he became the Director of the DSİ at the young age of thirty-one; he held this post for six years. While the Director of DSİ, he supported the Keban Dam project as critical to developing the Euphrates river basin and supervised the organization's ongoing planning and survey works. After a military coup d'état in 1960 against the Democrat Party (Demokrat Parti; DP) government, Demirel left DSİ and completed a period of military service working for the State Planning Organization (Devlet Planlama Teşkilatı; DPT); he participated in crafting Turkey's First Five-Year Development Plan. Upon returning to civilian life, he worked as a private consultant and, fittingly, a representative of the American engineering and construction firm, Morrison Knudsen, one of the six companies contracted to build the Hoover Dam. In 1964, a newly constituted Justice Party (Adalet Partisi; AP) elected Süleyman Demirel its chairman, and, on October 27, 1965, he became Turkey's youngest prime minister. With impeccable timing, the "King of Dams" arrived in office in time to lay the foundation stone of the Keban Dam seven months later. Demirel also rewarded engineers who had helped to shepherd the Keban project with cabinet positions in his new government.[6]

On the one hand, Demirel's biography reveals aspects of the creation of a technocratic state, where political legitimacy depended on the ability to demonstrate a supposedly neutral techno-economic prowess. Over the course of his career, Demirel's combination of technical knowledge and leadership provided a basis for the development of political power, expanding both his own authority and that of the applied sciences in public discourse. In part, this came about because technical expertise and development projects allowed the Turkish government to access a vital, geopolitical wellspring of economic and political support in the United States. The Cold War initiated a political competition between the two superpowers that spanned many realms, from nuclear weaponry to economic development (Westad, 2005; Latham, 2010; Hecht, 2011). An astute Ankara government could thus access a veritable fountain of development assistance, but only if it framed its applications properly. One convenient framing was Cold War competition; the case for building the Keban Dam was strengthened when Syria announced that it would also build a mega dam on the same river but with Soviet assistance (El-Khatib, 1984).

On the other hand, Demirel's biography is a story of the translation and contextualization of technical forms of knowledge into the language and methods of governance. As with the transformation of the metaphor of the Euphrates River, the evolution of Demirel's career from engineer to bureaucrat to premier demonstrates an investment of technical knowledge with new authority. The "knowing" of nature in particular ways and for specific purposes was hardly simply "management"; it was an orchestration of meanings with broad implications for sociopolitical life. Demirel's ascent from water engineer to prime minister helped further a wider trend, as governments once peopled by civil servants and military officers came to include an increasing number of individuals with technical backgrounds (Göle, 1993: 200). Indeed, learning how to translate technical knowledge into capital investments and connecting the whole to a concept of "civilization" was how political power was coming to work in the Turkish Republic.

All this talk of fish: the feasibility report and international finance

Demirel's biography and personal reflections tell us much about the connection of natural resource exploitation and governance in post-1945 Turkish history. His personal history does not, by itself, demonstrate the mechanics of technical knowledge, the way this type of knowledge production worked to make a dam. Whereas poetry by its nature offers itself up for interpretation, scientists and engineers intend their observations to be empirical and self-evident. Yet, when translating technical knowledge into society, a considerable amount of work must be done (Latour, 1987). In the case of the Keban Dam project, making technical designs into international finance required the production of a peculiar kind of document, the third-party feasibility report. A feasibility report outlines how a large-scale infrastructure project might work. As noted above, Timothy Mitchell has argued that a dam rearranges *both* the social and natural environment. The "feasibility report" is a geopolitical equivalent; it is the knowledge that arranges the political and financial institutions necessary to undertake changes to society and the environment.

Now, one could simply characterize the feasibility report a la James Scott (2008), as a document that "sees" the environment using methods of abstraction and representation. But that would leave historical analysis in the realm of general political theory, without the specificities of how this report did its work in the world. The Keban project's feasibility report attempted to do something in particular: connect the entire Turkish economy to the building of the Keban Dam. Such an effort was inherently political and far from self-evident. Specific historical contexts – the Turkish government's process of funding the report, its writing by American engineers, the World Bank's eventual acceptance – influenced the making of the feasibility report even as the report itself sought to reframe and remake social and environmental conditions in Turkey's east. Begüm Adalet has noted in recent work how much these historical contexts matter in tales of modernization. While modernizing reformers (and, at times, later analyses) often touted a triumphalist narrative, a careful detailing of the politics suggests that "fragilities and anxieties . . . mark expert thinking and practice" (2018: 14). This is no less the case here, as the battle to fund the feasibility report will demonstrate.

In 1961, a group of MPs in the Turkish parliament submitted a proposal to add five million Turkish lira to the budget of the EİE. The legislators intended the extra monies to fund a foreign firm in the preparation of a feasibility report; the report was the first step in obtaining credit from providers of foreign development aid. While a paltry sum in comparison to the overall cost of the dam, paying for such an effort signaled a larger political commitment to the Keban Dam project. The parliamentary Budget and Plan Committee, which considered the proposal first, collapsed into acrimonious debate. Resistance to the proposal was stiff. Fethi Çelikbaş, under whose Ministry of Industry the EİE operated, declared that the money was not required and that he would not use it, even if it were appropriated. Another member of the committee, Dr. Suphi Baykam, MP from Istanbul, resigned in protest. The proposal's opponents felt the money would be better spent elsewhere on smaller projects that could be more rapidly realized, rather than preparing a study for a project that could take a decade or more to build (M. Turgut, 1995: 38–40).

Despite the acrimony, the proposal passed the committee for consideration by parliament, but, as parliaments are wont to do, another committee was formed to study the first committee's proposal. The National Assembly charged this new committee with determining whether the Keban Dam project fulfilled the ideals of the recently promulgated Five-Year Development Plan. The architects of Turkey's post-coup constitution of 1960 expected the Development Plan to direct allocation of government funds and reduce political infighting. The struggle over the Keban Dam suggested that the Plan was accomplishing the former goal, if not the latter. For some who supported the dam, a massive project in eastern Anatolia was in fact a fulfillment of the Plan's aims. For others, the Keban project seemed likely to absorb far too much attention and government revenue that could be more profitably deployed in their own jurisdictions.

Attaching the feasibility report to the Development Plan's aims thus became the only way for the new committee to move forward with the project. The Plan's authors had forthrightly declared the Plan's overarching goal, "the achievement of a high growth rate and the consequent rise in incomes is not the ultimate aim. The real aim is to promote social welfare." Moreover, as noted above, the Plan specifically targeted the eastern half of the country for development in order to promote "the establishment of a balanced social order" (State Planning Organization, 1963: iii). To that end, the dam's proponents demonstrated all the ways the installation would fulfill this promise, while the project's detractors offered a number of other, less expensive options that could accomplish similar aims. After six months of debate, ultimately, the question was not whether to build a dam;

it was which dam to build first. There was little questioning of whether hydro projects were, in fact, vital to "social welfare." Though the representative of the DSİ, Senior Engineer Selahattin Kılıç, presented the committee with a number of different project options, a vision of a massive dam on the Euphrates River won out in the end (Turgut, 1995: 41–42). The committee issued a final report in support of the Keban Dam, in which it declared:

> We earnestly desire the benefit to our country of this project by its quick realization. This project will play a large role in the social and economic development of Southeastern and Eastern Anatolia and as large a role in agricultural and industrial growth as in the question of our country's energy.
> (Millet Meclisi Tutanak Dergisi, 1962: 359)

The commission recommended that the dam be included in the First Five-Year Development Plan and concluded that the Keban Dam was the "key project" for improving the social and economic conditions of this undeveloped part of the country (359). With this report, the Turkish parliament allocated funds for a feasibility report. In December 1962, EBASCO Services Inc., an American engineering firm headquartered in lower Manhattan, signed a contract to produce the report.

The very birth of the feasibility report was political, so it cannot be thought of as merely a neutral, technical document meant only for experts. Rather, the feasibility report functions as an artifact of a political process in which a massive effort to build the world's largest dam, creating Turkey's third-largest lake in the process, came to represent social reformation, which was put in a number of different ways: "a balanced social order" or "social welfare" or "the social and economic development of Southeastern and Eastern Anatolia." Funding the production of the report became so contentious because it committed *this* dam for *these* purposes, tying a particular project to a conception of how the eastern half of Anatolia was meant to change. Of course, that change could only come about once the feasibility report did its work: the marshaling of international aid. The authors of the feasibility report had specific methods to bring the dam into existence, just as the poets and writers who sought to alter the Euphrates metaphor, and the engineer-turned-prime-minister Süleyman Demirel, who sought to adapt features of American-style "civilization" to Turkey.

The EBASCO feasibility report connected a place, an environment, and a project to international finance. The feasibility report's authors also sought to demonstrate some relation to Turkey's overarching Development Plan in order to convince international funders that the dam embodied a larger transformation beyond the environmental and electrical. Most of the report understandably focuses on technical details such as the location of the dam and various choices of equipment; a large part of the report's 350 pages deals directly with the dam's composition, its location on the river, and the positioning of the electrical grid. Maps, schematics, and other statistics in these sections provide a range of technical details required by the engineer and planner. But the authors also departed from these matters and ranged widely, asserting a form of technocratic comprehensiveness.

The way the report's authors assert exhaustive scope reveals much about the Keban Dam project's true implications for Turkish society, economy, and politics. Clearly, a 350-page report cannot examine every facet, but certain choices stand out. For instance, the report's survey of electrical resources included "45 of Turkey's 67 provinces" and covered "78% of the total population" (EBASCO, 1963: III-1). However, of the forty-five provinces chosen by the authors for examination, only one, Diyarbakır, lies east of Keban. The engineers ignored the cities of Trabzon, Kars, Erzurum, Van, and Mardin, despite the parliamentary commission report asserting that the Keban Dam was "the key project" for the development of that part of the country. Of the fourteen provinces making up the Eastern Anatolia region, only two were part of the study area, while just half of the Southeastern Anatolia region was included. Moreover, the study area included only five of the seventeen provinces considered to form Northern Kurdistan (Culcasi, 2006). There was a clear reason for preferring an analysis of Turkey's western provinces. According to the report, nearly all of the electricity produced by the dam was to be sent west from Keban to Ankara and Istanbul to fuel factories, light homes and streets, and contribute to economic development in that part of the country (EBASCO, 1963: IV-24, map on IV-25).

While some Keban project electricity was meant for Urfa and Antep, the question arises whether the dam was really meant for the development of the Eastern and Southeastern Anatolia regions or for improving economic conditions in Turkey's largest cities in the west. There was no provision in the Keban project for agricultural development, for instance, despite the assertion of agricultural growth contained within the commission report to parliament.[7] According to the feasibility report, the "frontier" was not, in fact, to be re-engineered and incorporated – there was to be no recession or closure. In fact, the report offered a precedent: without a definitive end, the frontier idea might be repurposed again and again to support further technopolitical interventions without a clear detailing of the costs or benefits to local communities.

The report does pay some attention to overarching development goals, contending with various sectors of Turkey's economy such as mining, timber, and fishing. However, the details are odd. One might expect that these sections detailed the projected impact of the dam on these industries, perhaps through an enhanced provision of electricity. But that was not the case. Instead, EBASCO simply notes the location of Turkey's logging industry along the Black Sea coast some three hundred miles north of the Keban Dam site. With respect to the fishing industry, the authors point out that mackerel and Atlantic bonito, both saltwater fishes, represented seventy percent of the yearly catch for an expanding fish-canning industry based in Istanbul. What, one might ask, does the timber industry on the Black Sea or the fishing industry on the Sea of Marmara, both hundreds of miles from Keban, have to do with building a dam on the Euphrates River? Perhaps the new reservoir behind the dam could become a new locus for canning fish? That seemed unlikely, even to the engineers, who noted, "The fish-canning segment of the fishing industry is negligible, amounting to only 2 percent of total production" (Ibid.: II-7).

All this talk of fish points to an underlying logic, not only in the feasibility document, but also in the whole enterprise of building the Keban Dam. An infrastructure project of this magnitude required successive layers of knowledge production happening on different scales for different audiences: the poetics of local support, the political ascendance of technical expertise, the geopolitical maneuvering and massive capital outlays at the heart of "technical assistance." No single mode of knowledge was sufficient: each had its role to play, each required simultaneity. The layering of this knowledge production ultimately involved a kind of compounding; in a sense, the layers earned interest, in virtually all senses of that word. Once assembled, this layered knowledge became replicable and helped to build the sociotechnical "circuitry" necessary for future projects.

Conclusion: a 'project' for southeastern Anatolia

The Keban Dam's physical properties made the construction of additional dams on the Euphrates River more plausible because of its control of the river's flow. At the same time, the dam project, as a result of these concurrent layers of knowledge production, helped delineate the terms for the social and economic "development" of the southeastern Anatolia region. As a physical intervention in the environment and an industrial facility, the Keban Dam had a number of negative effects on the surrounding social and natural terrain. The dam's reservoir forced the relocation of nearly 40,000 people, inundated a number of important archaeological sites, and flooded roads connecting the Elazığ province to other regional centers. An important local employer in the province, a sugar factory, was located in the reservoir's footprint and had to be rebuilt elsewhere. But these important ramifications of dam construction are not the whole story. The intertwined processes of knowledge production detailed above tied the dam to notions of social and economic welfare, making the Keban Dam equally a sociopolitical intervention. This intervention was Janus-faced. On one side, the dam promised to resolve the "problems" of an eastern Anatolia "frontier" region that had occupied the minds of Ottoman and Turkish modernizers for generations. On the other side, the dam's combined effects helped to make southeastern Anatolia into a technopolitical object, in essence preserving the "frontier problem" for the future in order to justify continued technopolitical interventions, namely additional dams.

After the publication of the EBASCO feasibility report, engineer Korkut Özal, younger brother of Turgut Özal, began to envision a whole series of dams along the Euphrates River and its companion watercourse, the Tigris. The Özals were colleagues of Süleyman Demirel; older brother Turgut was appointed by Demirel to head the DPT where he gained significant power over the writing of the Second Five-Year Development Plan for the period 1968–1972. Turgut later became Prime Minister of Turkey from 1983 to 1989 and the country's eighth president from 1989 to 1993. His younger brother, Korkut, remained an engineer for much of his career. In 1951, he joined the DSİ, serving first at the branch office in Malatya, an important trading city located just west of the Euphrates River, and then as general director of the Ninth Regional Office in Elazığ, the provincial center just 45 km from Keban. After ten years in the bureaucracy, Korkut Özal took up an academic position at the Middle East Technical University. While there, he built upon his work at the DSİ and published a number of studies on the Keban Dam project and the development of the Euphrates river basin. Korkut entered politics in the 1970s, serving as Minister of Agriculture from 1974 to 1977 and briefly as Minister of the Interior.[8]

Korkut Özal's studies of the Euphrates River imagine a southeastern Anatolia radically reshaped by dams and irrigation projects. In *Keban Barajı ve Ötesi* ("The Keban Dam and After"), published in 1964 when Özal was working at the Middle East Technical University, he extols the dam as "the necessary first project" that would eventually open southeastern Anatolia to a larger transformation (10). Some aspects of this transformation are plainly explained in the report. While the Keban Dam initially would do little to expand irrigation, the downstream installations made possible by the Keban project could irrigate over 1.5 million hectares. This, Özal noted, was more than one-third of the total irrigable territory of the Turkish Republic. The Keban project and its "after-dams" would also produce billions of kilowatt-hours of electricity, perhaps useful for founding an aluminum or fertilizer industry (Ibid.: 15, 25). Other aspects of this transformation, though, were left unmentioned. For instance, the report does not discuss the social and environmental effects of huge reservoirs that would flood hundreds of square kilometers or the myriad health effects of irrigation, already well-known by the mid-1960s.

So, though Özal self-consciously writes about a particular natural formation, a river basin, the environment in his academic work is only what can be built; there is only the barest reflection on eastern Anatolia as a place where anyone or anything lives. Natural processes such as salination and silt are rendered only in relation to the dams themselves; the great installations, we are told, will "improve water quality" and "trap sand." Meanwhile, Özal expresses his concern for human society through a form of shorthand. Energy becomes industry. One must assume industry produces gainful employment and rising living standards. Water becomes irrigated fields, which, again, one must assume produces a social good. Indeed, the very fact of investment in these huge construction projects is taken as a self-evident cure for social problems, "the [Keban] plan, with its great investments in the Euphrates basin, offers great possibilities in terms of the social and economic issues that continue to prevail in this region" (Ibid.: 6). There is no further discussion in the sixty-page document about how such investments will translate into social and economic life.

Özal's academic work on the Keban Dam and "its importance to the entire Euphrates Basin," uses some forms of knowledge – irrigable land in hectares, electricity demand based on projected industrialization, etc. – to occlude other social and political processes. Justifying the dam required only certain kinds of ideas and connections – the rest could be assumed. Özal's text, *Keban Barajı ve Ötesi*, is in both title and form expressly about a future, having been published before the beginning of construction work on the Keban project. While saturated with charts and figures, it is useful to consider Özal's writing as an imagination, a work meant less to reflect reality than to produce a new one through the arrangement of certain kinds of knowledge and meanings. For Özal, the future comprised, in the main, what ought to be built in the river basin, either by Turkey or by the other states in the river basin, Syria and Iraq (26 and 56–57).

In Turkey, these future efforts would eventually combine into a single conception of change under the name, the Southeastern Anatolia Project (*Güneydoğu Anadolu Projesi*, GAP). Though sometimes seen as a water development scheme of the 1980s, conceiving of the GAP relied on knowledge production at multiple sites and on multiple scales throughout the late 1950s and 1960s. Özal's efforts combined with other discourses: the shifting of the public imagination of the Euphrates River, the legitimation and political authority obtained by engineering experts, and the linking of the Keban Dam project with ideas of social "balance" and welfare. Together, these ideas and practices underpinned and helped produce the Keban Dam, while also setting the boundaries, the political mechanisms, and the terms and meanings that substantiated a technopolitical concept of southeastern Anatolia.[9]

The Keban Dam may then be thought of as a technopolitical frontier in at least two, overlapping senses. First, there was eastern Anatolia as a historical frontier and object of reform, a concept already present in the late Ottoman period. Technopolitics requires a "problem" needing a technological "fix," so the dam's proponents revivified that earlier frontier idea within the vernacular of post-WWII modernization, recapitulating an "underdeveloped" eastern Anatolia that could only benefit from investment in infrastructure. It was not only physical infrastructure at stake, though. There was also cultural infrastructure to be built. In the sphere of metaphor, meaning, and landscape, the Euphrates River's life-giving freedom and wildness were reformed into an ambition to tame and harness the river's flow.

Yet, a closure of that historical frontier was not really the purpose of the Keban Dam project. Later history and the EBASCO feasibility report itself suggested the second sense of the technopolitical frontier: an enfolding of the historical frontier into a broader technopolitical framing of southeastern Anatolia. In this way, the Keban Dam ensured a kind of "path dependency," necessitating, on the one hand, nearly two dozen more dams and, on the other hand, the perpetuation and extension of a frontier concept, now wholly encased within a technopolitical frame. Ultimately, the Keban Dam's completion at a multifaceted, multilayered confluence of rivers, geology, culture, politics, and technical knowledge preserved a particular version of southeastern Anatolia as a "project," an object of unceasing technopolitical re-envisioning which continues to this day.

Notes

1 In relation to eastern Anatolia and the idea of frontier, consider Klein, 2011; M.A. Reynolds, 2011; Karpat and Zens, 2003; and Cora, Sipahi, and Derderian, 2016. There is also a wider literature in Ottoman environmental history that discusses attempts by

Ottoman officials to reform "frontier" regions of the empire. See, for example, Gratien, 2017; Low, 2015.
2 A fine discussion of Young Turk policies regarding both Christians and Kurdish Muslims in eastern Anatolia may be found in Üngör, 2011: 55–169. A broader analysis of the Ottomans' conception of their "civilizing mission" in these territories may be found in Deringil, 2003; Makdisi, 2002.
3 Though not detailed in this study, traces of resistance may also be found in the surveys conducted as part of the archaeological rescue work done in advance of the rise of the Keban Dam reservoir. See Dissard, 2011.
4 Turkish Muslims believe Urfa was once Ur of the Chaldeans and the birthplace of the prophet Abraham. For more, see Batuman, 2011.
5 These included the Demirköprü, Kemer, and Hirfanlı dams. Turgut, 1992: 118. For more on the Bureau of Reclamation, see Sneddon, 2015; Pisani, 2002; Worster, 1985.
6 Demirel's role in supporting the Keban project is noted in Turgut, 1995: 21. For his post-DSİ work and rise to power, see Komşuoğlu, 2008: 110–111; Arat, 2002: 88.
7 An irrigation water pumping project, Uluova Eyüpbağları, using Keban reservoir water was eventually constructed in 1987; another project, Kuzuova, opened in 1999 (Baş, 2016).
8 See Özal, 1963, 1964. Korkut Özal's tenure as Minister of Agriculture was interrupted for a period of roughly seven months between November 1974 and July 1975 by a change of government. He served as Minister of the Interior for six months from July 1977 to January 1978.
9 Geographers have also argued for cross-scale analysis of the Southeastern Anatolia Project, though these works retain a more contemporary focus. See, for instance, Harris, 2002.

References

Adalet, B. (2018). *Hotels and Highways: The Construction of Modernization Theory in Cold War Turkey*. Stanford, CA: Stanford University Press, 2018.
Akarun, R. (1999). *A Large Dam on Difficult Foundation: Keban Dam*. Ankara: Yapı Teknik Engineering and Consultancy Co.
Akçam, T. (2013). *The Young Turks' Crime against Humanity: The Armenian Genocide and Ethnic Cleansing in the Ottoman Empire*. Princeton: Princeton University Press.
Arat, Y. (2002). Süleyman Demirel: National Will and Beyond. In M. Heper and S. Sayari, eds, *Political Leaders and Democracy in Turkey*, 1st ed. Lanham: Lexington Books, pp. 87–106.
Asya, A. N. (1963). Fırat. *Elazığ Gazetesi*, p. 2.
Baş, Y. (2016). Elazığ İli, Sarıyakup Köyü Tarihi. *Bingöl University Social Sciences Institute Journal*, 6, pp. 133–135.
Batuman, E. (2011). The Sanctuary: The World's Oldest Temple and the Dawn of Civilization. *New Yorker*, p. 72.
Biswas, A. K. (ed.). (1994). *International Waters of the Middle East: From Euphrates-Tigris to Nile*, 1st ed. Bombay: Oxford University Press.
Burçak, B. (2008). Modernization, Science and Engineering in the Early Nineteenth Century Ottoman Empire. *Middle Eastern Studies*, 44(1), pp. 69–83.
Citino, N. (2008). The Ottoman Legacy in Cold War Modernization. *International Journal of Middle East Studies*, 40(4), 579–597.
Cora, Y. T., Sipahi, A., and Derderian, D. (eds.). (2016). *The Ottoman East in the Nineteenth Century: Societies, Identities and Politics*, 1st ed. New York: I. B. Tauris.

Coronil, F. (1997). *The Magical State: Nature, Money, and Modernity in Venezuela*. Chicago: University of Chicago Press.

Culcasi, K. (2006). Cartographically Constructing Kurdistan Within Geopolitical and Orientalist Discourses. *Political Geography*, 25(6), pp. 680–706.

Davis, D. K. (2007). *Resurrecting the Granary of Rome: Environmental History and French Colonial Expansion in North Africa*. Athens, OH: Ohio University Press.

Davis, D. K. and Burke III, E. (eds). (2011). *Environmental Imaginaries of the Middle East and North Africa*, 1st ed. Athens, OH: Ohio University Press.

Deringil, S. (2003). 'They Live in a State of Nomadism and Savagery': The Late Ottoman Empire and the Post-Colonial Debate. *Comparative Studies in Society and History*, 45(2), pp. 311–342.

Derr, J. (2011). Drafting a Map of Colonial Egypt: The 1902 Aswan Dam, Historical Imagination, and the Production of Agricultural Geography. In D. K. Davis and E. Burke III, eds, *Environmental Imaginaries of the Middle East and North Africa*, 1st ed. Athens, OH: Ohio University Press, pp. 136–157.

Dissard, L. (2011). Submerged Stories from the Sidelines of Archaeological Science: The History and Politics of the Keban Dam Rescue Project (1967–1975) in Eastern Turkey. Ph.D. Berkeley: University of California.

Dökmeci, C. (1966). Fırat'la Söyleşme. *Yeni Fırat*, 28, p. 7.

EBASCO. (1963). *Report to Electric Power Resources Survey and Development Administration for Engineering and Economic Feasibility of Keban Dam and Hydroelectric Project*. New York: EBASCO Services Inc.

El-Khatib, M. (1984). The Syrian Tabqa Dam: Its Development and Impact. *The Geographical Bulletin*, 26, pp. 19–28.

Gençosmanoğlu, Y. N. (1963). Fırat'la Hesaplaşma. *Elazığ Gazetesi*, p. 2.

Göle, N. (1993). Engineers: 'Technocratic Democracy.' In M. Heper, A. Öncü, and H. Kramer, eds, *Turkey and the West: Changing Political and Cultural Identities*, 1st ed. London: I. B. Tauris, pp. 199–218.

Gratien, C. (2017). The Ottoman Quagmire: Malaria, Swamps, and Settlement in the Late Ottoman Mediterranean. *International Journal of Middle East Studies*, 49(4), pp. 583–604.

Hanioğlu, M. Ş. (2005). "Blueprints for a Future Society: Late Ottoman Materialists on Science, Religion, and Art." In *Late Ottoman Society: The Intellectual Legacy*, edited by Elisabeth Özdalga, 28–116. New York: Routledge, 2005.

———. (2017). *Atatürk: An Intellectual Biography*. Princeton: Princeton University Press.

Harris, L. M. (2002). Water and Conflict Geographies of the Southeastern Anatolia Project. *Society & Natural Resources*, 15(8), pp. 743–759.

Hecht, G. (1998). *The Radiance of France: Nuclear Power and National Identity After World War II*. Inside Technology. Cambridge, MA: MIT Press.

———. (ed.). (2011). *Entangled Geographies: Empire and Technopolitics in the Global Cold War*, 1st ed. Cambridge, MA: MIT Press.

Hommes, L., Boelens, R., and Maat, H. (2016). Contested Hydrosocial Territories and Disputed Water Governance: Struggles and Competing Claims over the Ilisu Dam Development in Southeastern Turkey. *Geoforum*, 71, pp. 9–20.

Karpat, K. H. and Zens, R. W. (eds. (2003). *Ottoman Borderlands: Issues, Personalities, and Political Changes*, 1st ed. Madison, WI: Center of Turkish Studies, University of Wisconsin.

Klein, J. (2011). *The Margins of Empire: Kurdish Militias in the Ottoman Tribal Zone*. Stanford, CA: Stanford University Press.

Kolars, J. F. and Mitchell, W. (1991). *The Euphrates River and the Southeast Anatolia Development Project*. Carbondale, IL: Southern Illinois University Press.

Komşuoğlu, A. (2008). *Türkiye Siyasetinde Bir Lider: Süleyman Demirel*. Istanbul: Bengi Yayınları.

Latham, M. E. (2010). *The Right Kind of Revolution: Modernization, Development, and U.S. Foreign Policy from the Cold War to the Present*. Ithaca: Cornell University Press.

Latour, Bruno. (1987). *Science in Action: How to Follow Scientists and Engineers Through Society*. Cambridge, MA: Harvard University Press.

Low, M. C. (2015). Ottoman Infrastructures of the Saudi Hydro-State: The Technopolitics of Pilgrimage and Potable Water in the Hijaz. *Comparative Studies in Society and History*, 57, pp. 942–974.

Makdisi, U. (2002). Ottoman Orientalism. *The American Historical Review*, 107(3), pp. 768–796.

McDowall, D. (1997). *A Modern History of the Kurds*. London: I. B. Tauris.

Millet Meclisi Tutanak Dergisi. (1962). *Keban Barajı ve Aşağı Fırat Havzası Kalkınma Projesi Hakkında Millet Meclisi Adına Araştırma Komisyonu*. Ankara: Millet Meclisi.

Mitchell, Timothy. (2002). *Rule of Experts: Egypt, Techno-Politics, Modernity*. Berkeley: University of California Press.

———. (2011). *Carbon Democracy: Political Power in the Age of Oil*. New York: Verso.

———. (2014). Introduction: Life of Infrastructure. *Comparative Studies of South Asia, Africa and the Middle East*, 34(3), pp. 437–439.

Öktem, K. (2002). When Dams Are Built on Shaky Grounds: Policy Choice and Social Performance of Hydro-Project Based Development in Turkey. *Erdkunde*, 56(3), pp. 310–325.

Olson, R. W. (1991). *The Emergence of Kurdish Nationalism and the Sheikh Said Rebellion, 1880–1925*. Austin: University of Texas Press.

Orhon, M., Esendal, S., and Kazak, M. A. (1991). *Türkiye'deki Barajlar/Dams in Turkey*. Ankara: Devlet Su İşleri Genel Müdürlüğü.

Özal, K. (1962). *Power Development Possibilities in the Upper Euphrates Basin Upstream of Keban*. Ankara: Enerji ve Tabii Kaynaklar Bakanlığı.

———. (1963). *The Report on the Hydroelectric Development Possibilities in the Lower Euphrates Basin in Turkey*. Ankara: Elektrik İşleri Etüt İdaresi Genel Müdürlüğü.

———. (1964). *Keban Barajı ve Ötesi*. Ankara: ODTÜ Mühendislik Fakültesi.

Pisani, D. J. (2002). *Water and American Government: The Reclamation Bureau, National Water Policy, and the West, 1902–1935*. Berkeley, CA: University of California Press.

Pritchard, S. B. (2012). From Hydroimperialism to Hydrocapitalism: 'French' Hydraulics in France, North Africa, and Beyond. *Social Studies of Science*, 42(4), pp. 591–615.

Reynolds, M. A. (2011). *Shattering Empires: The Clash and Collapse of the Ottoman and Russian Empires, 1908–1918*. New York: Cambridge University Press.

Reynolds, N. Y. (2012). Building the Past: Rockscapes and the Aswan High Dam in Egypt. In A. Mikhail, ed., *Water on Sand: Environmental Histories of the Middle East and North Africa*, 1st ed. New York: Oxford University Press, pp. 181–205.

———. (2017). City of the High Dam: Aswan and the Promise of Postcolonialism in Egypt. *City & Society*, 29(1), pp. 213–235.

Rogers, P. P. and Lydon, P., eds. (1996). *Water in the Arab World: Perspectives and Prognoses*, 1st ed. Cairo: American University in Cairo Press.

Scott, J. C. (2008). *Seeing Like a State: How Certain Schemes to Improve the Human Condition Have Failed*. New Haven, CT: Yale University Press.

Senemoğlu, B. (1963). Fırat ve Baraj. *Elazığ Gazetesi*, p. 1.

Sneddon, C. (2015). *Concrete Revolution: Large Dams, Cold War Geopolitics, and the US Bureau of Reclamation*. Chicago: University of Chicago Press.

State Planning Organization. (1963). *First Five-Year Development Plan 1963–1967*. Ankara: State Planning Organization.

Suny, R. G., Naimark, N. M., and Göçek, F. M., eds. (2011). *A Question of Genocide: Armenians and Turks at the End of the Ottoman Empire*, 1st ed. New York: Oxford University Press.

Turgut, H. (1992). *Demirel'in Dünyası*. Istanbul: ABC Basın Ajansı.

———. (2000). *GAP ve Demirel: 50 yıl*. Istanbul: ABC Basın Ajansı.

Turgut, M. (1995). *Gap'ın Sahipleri*. Istanbul: Boğaziçi Yayınları.

Üngör, U. Ü. (2011). *The Making of Modern Turkey: Nation and State in Eastern Anatolia, 1913–1950*. Oxford: Oxford University Press.

Westad, O. A. (2005). *The Global Cold War: Third World Interventions and the Making of Our Times*. New York: Cambridge University Press.

Worster, D. (1985). *Rivers of Empire: Water, Aridity, and the Growth of the American West*. New York: Pantheon Books.

Yeni Fırat. (1962). *Yeni Fırat*, 1(1), p. 1.

Zürcher, E. (1992). The Ottoman Legacy of the Turkish Republic: An Attempt at a New Periodization. *Die Welt Des Islams*, 32(2), 237–253. doi:10.2307/1570835

4 From imperial frontier to national heartland

Environmental history of Turkey's nation-building in its European province of Thrace, 1920–1940

Eda Acara

Marxist critic Raymond Williams draws attention to three cyclical meanings of nature: "(i) *the essential quality and character of something; (ii) the inherent force which directs either the world or human beings or both; (iii) the material world itself, taken as including or not including human beings*" (Williams, 1983 [1976]: 219).[1] These three meanings of nature are blended to strengthen constructions of what and where is called nature and natural. Through this complex articulation of meanings, this way of representing nature can made normal(ized) and/or normative by society (Castree and Braun, 2001). Similarly, in grappling with the complex meanings of nature, global environmental history has been interested in the ways that nation and nature are articulated, and in how some of these articulations become part of government policies and nation-building processes, which often have material consequences for human-environment relations (Radkau, 2008).

In this chapter, I am interested in exploring the meanings and representations of places in which nature and nation became politicized and entrenched in the fabric of the state's governing discourses in its nation-building processes during Turkey's early Republican era. I tackle the idea of nature to historicize how nature and modernity were constructed within the context of Turkish nationalism. I look at, for example, how factories were seen as a way to "tame nature" and how they were intended to generate urban development as a symbol of Turkish modernity. I explore state-space relations by closely examining state practices of territorializing Turkey's Thrace border region into part of a "modern", "urban", "Turkish" nation-state during the early Republican era (1920– 1940). I then contextualize the discourses around nature in terms of the governance and regulation of Muslim-Balkan immigration from the Balkans to Thrace from the early Republican era to the 1950s in order to territorialize the new Turkish nation by the region's agricultural and immigration policies.

The spatial transformation of the Ottoman Empire into a nation-state was mainly characterized by "assimilation of rural elements in urban environments and the use of urban models in rural settings" (Baydar Nalbantoğlu, 1997: 200). Urbanist Tarık Şengül (2003, 2009) refers to this territorial re-configuration as the "urbanization of the nation-state". This phase aimed to eradicate the contradictions embedded in the "Ottoman urban pattern – lack of a centralised politico-spatial system, the organisation of cities along ethnic lines and the organic structure of cities that prevented the development of a state-centred control and surveillance system" (Şengül, 2003: 156). Other urbanists, like İlhan Tekeli (2013) and Çağatay Keskinok (2010), similarly draw attention to compulsory urban development plans in this era for all settlements with a population over 2,000 while the establishment of municipalities across Anatolia was another new spatial configuration within the Turkish nation-state territory.[2] In contributing to this literature, my argument is that the territory making of the early Republican era aimed to eradicate the Ottoman imperial past. This involved the construction of ideal 'modern' villages and state-factory settlements and advertising them in state publications as ideal places within the Turkish nation-state. Here, nature was either narrated as something to be modernized or depicted as the scene of a disastrous Ottoman past. Moreover, nature was imagined within a context of historical re-narration as a socio-spatial project for distinguishing Turkish national origins from the Ottoman Empire. This process of re-inscribing was possible by a new spatial configuration shaped by the urbanization of the nation-state. Thus, municipalities and urban development plans were not only the means to producing territory.

The early Republican Turkish state relied heavily on infrastructure building, which was planned and executed in Thrace by a regional coordination and command institution, *Trakya Umum Müfettişliği* [Public Inspectorship of Thrace]. In examining its operations, I conducted a discourse analysis of *Trakya Dergisi* [*Thrace Journal*], published monthly by the inspectorship itself. Its aims, to educate the Turkish peasantry about agriculture, diseases, and the Republican revolution, provided the complementary units of the analysis. The rest of the empirical material on which this chapter rests on is based on further analysis of oral history conducted in Alpullu Sugar Factory in 2012.

I conducted interviews with six male former workers. Five were between 50 and 60 years old while one was 86, and had worked as the only photographer in the factory for more than 20 years. The semi-structured, exploratory interviews, which lasted approximately one hour, focused on their memories of Alpullu Sugar Factory, its contributions to the town, and the effects of pollution of the Ergene River on sugar beet production. Another crucial theme concerned the on-going privatization of Turkey's sugar factories. The interviews were carried out in 2012 for my dissertation project, which focused on an entirely different topic: contemporary regional conflicts due to water pollution. Consequently, they only provide supplementary material for the analysis here due to their limited scope.

The methodology of this study depends on discourse analysis. By discourse, Michel Foucault meant "practices that systematically form the object of which they speak" (Foucault, 1974: 49). Discourse is about knowledge production through language and knowledge production is not merely a linguistic process, but a process related with social practices (Hall, 2001b). In parallel, according to Fairclough (2003), discourses manifest themselves through texts whose social influence is based on meaning-making processes. These are composed of three elements: 1) text production; 2) the text itself; 3) text reception (Fairclough,

2003). Drawing on Norman Fairclough's understanding on the co-constitution of text and context (1992, 2003), I apply my analysis to the early Republican context, which influenced the stages of production and reception of the journals in question. In the second step of my analysis, I content analyze *Trakya* by organizing the themes from the texts around the spatialization of the nation in daily life places. As a government-sponsored journal, *Trakya* represents the era's official nationalist discourse and thereby reflects the imagined nation and its places. In the final step, I blend in the oral history interviews to understand how territorial tactics are remembered and re-iterated today.

The chapter consists of two main sections. The first outlines the theoretical and historical context in which the empirical material of the research is analyzed. The second section presents the main argument by looking more closely at the constructions of nature, and the eradication of Ottoman pastoral meanings of villages and factories for the Turkification of the regional territory.

National territory of regional practice

Territory is "more than merely land, but a rendering of the emergent concept of 'space', as a political category: owned, distributed, mapped, calculated, bordered, and controlled" (Elden, 2007: 578; also see Elden, 2010). Territory encompasses not only the physical borders of the nation-state (Agnew, 1994); its boundaries also stretch over imaginary places through which "the territory exists and achieves institutionalized meanings" (Paasi, 2003: 113). Territory therefore "has to be worked for", which translates into techniques and practices to territorialize (Painter, 2010: 1105). The majority of the literature on territorialization focuses on for example, the state's legislative mechanisms, calculative institutions and/or accounting practices to deepen our understandings of how territories are produced as bounded, congruous, and coherent (Painter, 2010). Raffestin and Butler (2012) emphasize that "territories are produced by means of territories", which corresponds to reconstructing, re-using, and re-appropriating – in other words, re-territorializing – an existing territory through material and representational strategies. These strategies dynamically produce the permanence of territorial practice as belonging to states. These territory-producing mechanisms and processes also make history (Paasi, 2003; Brenner, 2004; Elden, 2013). Finally, produced territories are usually citational, where citationality refers to a form of continuation in material, social, institutional, and/or representational ways that was inherited from the former territoriality (Hakli, 2008; Raffestin and Butler, 2012).

The act of re-territorializing in this case required local and national elites to create and impose a new sense of territoriality from a multi-ethnic empire to a nation-state. This spatial process was built on a combination of past (imperial) and present (national) economic and social policies, re-shaping settlement, housing, and spheres of agricultural and industrial production. Thrace provides a rather important case for observing the full extent of these policies, as it was a major receiver of incoming Muslim Balkan immigrants from the Balkan

Wars (1912–1915) until the 1990s. While there were several immigration waves of Muslim-Balkan immigrants into Turkey from 1915 onwards, the largest movement occurred in 1915 following the Balkan Wars. Another occurred just after the Republic was established, when a population exchange treaty was implemented between Greece, Bulgaria, and Turkey. The latest occurred in 1950 and 1989, mainly from former Soviet countries due to ethnic tensions (Parla, 2005). The former local populations, who were mainly Jewish, Greek, and Bulgarian, and lived as Ottoman citizens, were exchanged and/or violently expelled from Thrace while Muslim-Balkan immigrants were resettled in former Greek or Bulgarian settlement zones (Cagaptay, 2003; Karpat, 2012). Maps provided by Şükrü Aslan et al. (2014, see especially 179–186) showing changes in mother tongue, demonstrate the significant decrease in non-Muslim populations, such as Bulgarian, Greek, Ladino, and Pomaks, between the 1927 and 1960 censuses.

The demographic shift after the Balkan wars and the First World War meant that the region's population, encompassing Tekirdağ, Kırklareli, Edirne, and Çanakkale, was around 700,000. Given that 190,000 of these people resided in provincial centers, population density was higher in the interior than coastal zones (Burgaç, 2013). Repopulating the region was a state territorial goal, which was to be achieved by establishing racialized ethnic hierarchies. Soner Cagaptay (2006) argues that the Turkish government saw Muslim-Balkan immigrants as the most favorable ethnic community for assimilation and Turkification. Accordingly, they were dispersed among Turkish inhabitants to enable their assimilation (Ibid.). In this context, the settlement law was a piece of governmental technology: an important social engineering tool from the 1930s to 1960s for regulating the resettlement and de-settlement of different ethnic populations across Turkey.

Since 1923, the Turkish state has framed the Muslim-Balkan immigration under an umbrella term, *muhacir*, a word derived in the early twentieth century, when the first wave of immigrants started to migrate to Turkey after the loss of Ottoman lands in the Balkans due to the Balkan Wars (1912–1915) (Kale, 2014). While these immigrants were initially resettled along the Ottoman Empire's European frontier as a demographic and military measure, they started to move back into Turkey after the Empire's demise. Thus, Thrace became a transitory gateway or border region while also maintaining political and territorial significance until the end of the Second World War due to increasing competition between Greece, Bulgaria, and Turkey over nationalist land claims (Ahmad, 1993).

Contextualizing public inspectorships

Public Inspectorships (*umum müfettişlikleri*) were military institutions that aimed to ensure regional security and economic development. They were first established sporadically during the late Ottoman Empire. The first Public Inspectorship, known as *Anadolu Islahatı Umum Müfettişliği* or *Anadolu Vilayet-i Umum Müfettişliği* (Public Inspectorship of Anatolia) (1895–1899) was established in the Empire's eastern provinces (specifically Erzurum, Sivas, Van, Bitlis, Elazığ, Diyarbakır, and Sivas) to suppress armed ethnic conflict in the region, implement reforms, and carry out institutional and other infrastructural modernization projects (Koçak, 2003; Burgaç, 2013).[3]

During the early twentieth century, the Public Inspectorship became increasingly important due to nationalist revolts in Macedonia (Salonica, Kosovo, and Manastır) (Adanır, 1996(1979); Koçak, 2003). The Turkish Republic then inherited these institutions from the Ottoman Empire. Like its predecessor, it used them as a territorial security tactic to counter armed unrest particularly in Kurdish territories. These institutions further aimed to initiate local development, to provide territorial economic security (Koçak, 2003; Burgaç, 2013).[4] Five public inspectorships were established in Turkey between 1927 and 1947.[5] In 1936, the Public Inspectorship of Thrace was the second to be established. It had the power to declare martial law, and all security forces were duty bound to it. Another mission was auditing the operations of government officials to hinder any form of illegal activity. Lastly, the inspectorship was designed as an institution to promote regional economic development (Koçak, 2003). The majority of Public Inspectors were recruited from military officers to maintain regional security and lead and/or implement modernization reforms. During the early Republican era, these two territorial missions sought further to dismantle the Ottoman Empire's provincial system in order to nationalize the regional economies (Koçak, 2003; Burgaç, 2013). Public inspectorships were thus inherited institutions, from imperial provincial rule to a centralized nation-state.

The Public Inspectorship of Thrace was an institution of power in a productive sense (Foucault, 1989). That is, power is not always repressive but "traverses and produces things, . . . forms of knowledge, produce[s] discourse" (Hall, 2001a: 77). Thus, the Public Inspectorship's territorial practices sought to produce space by strategizing territory and territorial practices as a political technology (Jones, 2007; De Carvalho, 2016). That is, they wished to produce the nation-state territory by cultivating the land as national land and its people as national subjects. The territorial practices consisted of bordering and cultivating land in addition to managing class and ethnic boundaries to produce the nation-state territory. This political process was achieved through a network of territorial relations that "have social space (spaces) for support" (Lefebvre, 1976–1978, vol. 4: 164–165 cited in Brenner and Elden, 2009). Thus, I would argue, the ideal villages and factory settlements were constructed by the inspectorships to 'make territory'. These establishments were seen by the Republican elites as symbols of the Turkish nation-state and Turkish nationalism while also signaling the territorial rupture from the decaying imperial past.

The primary tactics for governing early Republican spatial politics involved a discourse of Turkish migration alongside the enforcement of the Settlement Law (*İskan Kanunu*, law no. 2510). Turkish national identity was discursively produced with its own connotations of Turkish territoriality by what I term "the discourse of Turkish migration". The state's instrumental and strategic use of anthropometrics to institutionalize Turkey's territory and its national population defined the Turkish race, who had been migrating for centuries from Central Asia, as the historical agents of civilization, and assigned Anatolia as the (final) home of the Turkish nation (Kadıoğlu, 1998; Altınay, 2004). Although this narrative superficially recognized differences among Turkey's various ethnic populations, their names were not enunciated[6] but erased through a discourse that emphasized Turkish origins and a common history. This discourse of Turkish migration was very influential in the formation of the *Turkish History Thesis*, on which the content of the era's new geography books was based.[7]

While the discourse of Turkish migration cultivated the "imagined community" as a Turkish nation (Anderson, 1991), the Settlement Law and its implementation by the Public Inspectorship of Thrace helped to secure Thrace as part of Turkish national territory. On the other hand, the modernization of villages and peasantry, and the promotion of factory sites as modern urban places of the Turkish nation were territorial tactics of the emergent nation-state. In the next section, I illustrate these (central) state tactics to regulate and rule the emerging polity.

The eye of the nation-state against unruly distances

As "both a goal to achieve statehood and a belief in collective commonality" (Anderson, 1991; Nagel, 1998: 247), nationalism is more than a political ideology. It is a constitutive discourse that governs everyday relationships by reordering and governing the family, the relationships of family members with one another, and the relationship of families with each other and other institutions (Sirman, 2007). It is in this sense that nationalism aims to rule and reorder the distance by which the territoriality of the nation is made and becomes a state tactic. Accordingly, during the early Republican era, the Public Inspectorship of Thrace aimed at establishing regional authority to orchestrate the distance between the central government and Thrace as a border region and the distance between newly settled immigrants and the Republican regime. This continued regional distance, which hindered complete authoritative centralization, was a phenomenon inherited from the Ottoman Empire.[8]

Ruling national distances nationally and administratively during the early Republican era was an unfinished project that required territorial means. This was achieved firstly by turning the new Republic's land into a (Turkish) homeland (*vatanlaştırma*)[9] through settlement policies, among other reforms,[10] and secondly by re-establishing administrative authority between the border regions and the central government. The Settlement Law was a significant practice in the first step for making Thrace a homeland specifically for Muslim-Balkan immigrants while also dispossessing and displacing ethnic populations in other regions, specifically in Kurdish territories. Thus, the Settlement Law involved determining ethnic hierarchies on a territorial scale as the law privileged incoming Muslim-Balkan immigrants, except those with a Roma background, by ensuring their safe passage to Turkey and providing them with land. In the second step, the public administration structure was transformed, with the former imperial provincial system (*vilayet sistemi*) replaced by a city-centered system. This created a dual system where, for example, Thrace legally included four cities (Tekirdağ,

Kırklareli, Edirne, and Çanakkale), although practically they were administered as if part of Edirne province, just as in the Ottoman Empire (Burgaç, 2013). This was largely because the early Republican era was a transition from an imperial administrative economic geography to a nationalist one. Initially, therefore, the Public Inspectorship of Thrace was considered to have re-established distance to exercise ruling power between the central government, regions, and cities. Like the other public inspectorships, it was not, however, a decentralized and local administrative unit but the field organization of a centralized government (Burgaç, 2013: 194). Thus, distances were largely the concern of the central authority.

The course of urbanism in Thrace

The Settlement Law was used to regulate, settle, and assimilate immigrant populations between 1923 and 1990 (Sosyal Planlama Başkanlığı, 1990), with the majority being settled in Marmara, Thrace, and the Aegean (Kirişçi, 2000; Parla, 2005; Cagaptay, 2006).[11] For example, of 174,280 immigrants that entered Turkey between 1934 and 1947, 87,529 were settled in Thrace (Burgaç, 2013: 203). The Settlement Law was both a social and a geographical engineering project. Accordingly, the country was organized into three different zones. The first covered places where the central government wanted to increase the Turkish demographic presence. The second included places designated for settling immigrants, specifically the Thracian counties of Tekirdağ, Edirne, and Kırklareli. This zone also aimed to assimilate the immigrants into Turkish national culture and society (Hür, 2011). The last zone comprised areas where settlement was restricted because of health, economic, cultural, and/or political insecurity. Journalist and historian Ayşe Hür (2011) notes that the third zone involved politically rebellious sites located in the eastern and southeastern Turkey, where non-Turkish groups lived,[12] such as Ağrı, Dersim,[13] Kars, Bingöl, Bitlis, Muş, Batman, and Diyarbakır. Once immigrants were settled in a zone, they were forbidden to settle in another place for at least ten years (Law 2510, Article 29). The Settlement Law was thus both a re-territorializing and de-territorializing practice, as it established ethnic hierarchies within and across different regions.

In Thrace, many Muslim-Balkan immigrants were settled in former Bulgarian villages that were emptied after the Balkan Wars. Their deserted land needed infrastructural work for cultivation (Burgaç, 2013), so one of the most significant tasks of the Public Inspectorship of Thrace was preparing the land and other settlement conditions for constructing homes, including providing land for the immigrants, as the goal was to settle them and make them productive. By 1937, 15,000 immigrant houses had been built and 12 immigrant villages (seven in Tekirdağ and five in Çanakkale) had been constructed, while 78,000 immigrants had been settled (Burgaç, 2013). The Public Inspectorship of Thrace constructed water infrastructure in Thracian villages and planted small forests close to each village.

Eradication of the Ottoman past and Ottoman nature

The Turkish nation-space in the early Republican elite imaginary was dominated by secularism, rationalism, and scientific social engineering, which aimed
to erase the Ottoman imperial past. A nationalist narrative of "rupture" from
the Ottoman imperial past was established and disseminated by the national
and local press. This was used to legitimize numerous legal, educational, and
economic reforms and other infrastructural interventions by the Turkish Republic (Ahmad, 1993; Karaömerlioğlu, 1999; Akşit, 2005; Ahmad, 2008) because
the rupture was intended to break from a decaying past temporally and spatially.
Historian Elif Akşit (2010) refers to the "politics of decay" as a form of exclusion
that turns the decaying part into an object to demolish and re-construct. For
example, one short story in *Trakya* depicts the Ottoman Empire as the cause of
ruin in Anatolia. The story involves a fictional conversation between an Ottoman Sultan and Karagöz, a figure in traditional shadow play during the Ottoman period who gave voice to the illiterate public. While walking together in
Istanbul, they reach some ruins where two owls are talking to each other. When
the Sultan asks whether Karagöz can understand their language, he responds by
translating their conversation:

> "The female owl responds to the male owl by saying, 'I have to make sure of
> your love so that you create ruins for me to land on and hoot.'
> Well . . . what does the male bird respond?
> Karagöz listens to the owls for a while and responds as follows:
> ". . . The male bird continues: 'As long as an Ottoman Sultan rules this
> land, it is ruins that you are asking for'".
>
> (Anonymous, 1936a)

Erasing the Ottoman past, in other words cleansing the ruins of the past, was
the main objective during the early Republican era (Ahmad, 1993). The new
capital city, Ankara, was seen by the elites as an "expression of its [Republic's]
desire to create a new culture and civilization on the ruins of a decadent imperial past" (Ahmad, 1993: 91). Based on articles in *Trakya*, this eradication of the
Ottoman past was carried out in the Thrace region by reinventing the meaning of
the Turkish village, peasantry, and industry as places of the new Turkish civilization and nation.

The transformation of the meaning of villages unfolded in an authoritative
discourse of rural design. Villages were depicted as places embedded in the history of the Turkish race while the origins of Turkishness were nested in the place
of the villages. One of the most obvious examples of this understanding was
expressed in an article in *Trakya* in which Kandemir (1936) tells the story of
Turkish nomads who had migrated from Middle Asia to settle in Anatolia. Their
original name for their settlement, *köz* ["fire"], was gradually transformed into *köy*
["village"] over decades. This story reinforced the idea that the village, because

it was embedded in the history of the Turkish race, thereby formed the ultimate core of the Turkish nation:

> A long time passed between. All the Turkish groups made fires and settled on the land. As time passed, *köz* [fire] became *köy* [village] and each family took the fire into their houses and forged the village in their hearths.
>
> Different families, different lineages and different leaders, independent economic and moral powers all settled in the village. . . . Now the village lives on in Turkish history as a symbol of settled and conjoined powers.
>
> (Kandemir, 1936: 5)

The peasantry and village place were also depicted as fundamental units of national development and progress, and as a planning unit. There are a number of scholarly works about the ideology and activities of the People's Houses (*halk evleri*) in Turkey, concentrating on the nationalist and authoritative discourses that shaped the understanding of peasantist ideology (Karaömerlioğlu, 1999; Şimşek, 2002). This body of research argues that peasantist ideology circulated by the People's Houses targeted creation of a common national consciousness by decreasing the gap between the urban elites and the rural masses through ideological interventions at several spheres of culture, the social, and the political in the countryside. The interventions were shaped upon disseminating the Kemalist principles and ideals that were barely known by the peasants (Karaömerlioğlu, 1999).

A majority of the articles in *Trakya* emphasized the role of the village and peasantry in reconstructing the nation, which had been long neglected under Ottoman rule. For example, Öney (1936) cites a speech of the executive secretary of the ruling *Cumhuriyet Halk Partisi* (CHP; Republican People's Party) emphasizing the importance given to developing villages and their role in the national economy:

> The village is the source of food and raw products.
> The village is the largest exporter.
> The village is the largest and the most trustworthy client.
> The village is the basis of our army.
> The village in short means Turkey.
>
> (Öney, 1936: 15)

This understanding of the village was built on the notion of national unity, which bound the peasants both to each other and the government. Many *Trakya* articles emphasized the government's role in providing necessary funding and equipment to reconstruct old villages, restore their infrastructure, and improve agricultural development. These articles further stressed the peasants' responsibility to contribute to implementing the development plans in their villages; thus, one of the most common themes concerned the necessity for peasants to "work hard, together" (see Dirlik, 1936a; Anonymous, 1936b, 1936c, 1937; Kandemir, 1937; Öney, 1937). In these articles, the village symbolized a sense of place and belonging in Turkey's emerging nation-state. As a "family", the village stood at the core of building this sense of relatedness that all nationalisms aim to reach:

> In the past, the villages would work for the chief, now there is cooperation; everybody has their own property. This means that village issues will not be resolved according to the will of the chief . . . but they will be carried with our mind, knowledge and cooperation. The village is no longer a stranger to us. It is our own family.
>
> (Anonymous, 1936b: 4)

This "family" metaphor further reminds peasants of their responsibility for development, framed around love for the nation and land (for example, see Dirlik, 1936b). Feminist anthropologist Sirman (2007) and feminist historian Najmabadi (1997) argue that nationalism is built on scales of love that holds control over when, where, how, and between whom the act of loving will take place. In the early Republican project of nation-building in Turkey, these definitions were framed by discourses of development, which located the village and peasantry at its core.

Development was further depicted by nationalist elites as a tool against the imperial West, a tool for responding to colonial imperialism, and a way to erase the Ottoman past (Ahmad, 1993; Baydar Nalbantoğlu, 1997; Bozdoğan, 2001; Jongerden, 2009). With this objective, development remained at the core of nation-building and constructing a sense of nation-place and territoriality in Turkey, which ultimately led to the governmental projects to rehabilitate villages and their environments.

The village was seen by many of the contributors in *Trakya* as an abstract concept, historically familiar to members of the Turkish race (through the numerous civilizations they had built in and by their villages), but from which they had been distracted by Ottoman rule and imperial exploitation. Thus, modern development was seen as the only strategy to erase this past to build a different independent future. The idea of development therefore progressed alongside cleansing villages of their Ottoman past through a governmental war against nature, represented as a war against Ottoman neglect of villages. Accordingly, writers in *Trakya* represented the ideal village and its environment as places with drained marshlands, parks, urban networks, with clean water, electricity, roads, and modern concrete homes (Öney, 1936, 1937; Dirlik, 1936b; Anonymous, 1936c). One of the areas where such ideas were put into practice was the Public Inspectorship's program in Thrace to transform settlements into ideal villages. By 1940, it had created 250 ideal villages, representing 25 percent of all villages in Thrace and Çanakkale (Burgaç, 2013). According to Sayar (1952), Istanbul and Thrace perhaps had the most active spending for the construction of the ideal village and/or immigrant villages/neighborhoods as ideal villages since they had the most immigrants to settle.

For this period, whereas villages symbolized a rebuilt and reformed Ottoman past, the factory settlements and the cities and/or towns were the nation's future.

Industry was an ideological tool to save the Turkish Republic from economic exploitation and colonial dependency (Ahmad, 1993; Baydar Nalbantoğlu, 1997; Bozdoğan, 2001; Jongerden, 2009). Dramatic discourses around illness further helped reinforce the idea of national development. For example, anti-malaria campaigns were conveyed through antagonisms between modern versus traditional and urban versus rural, such that "the state's mission entailed . . . to cure malaria the disease but also to enlighten (or cure as a civilization) the peasantry as a class" (Evered and Evered, 2012: 314). Here, malaria-producing nature also stood for the decaying imperial past, with its connotations of underdevelopment and the peasantry. An article in *Trakya*, for example, emphasized how the town of Alpullu (in Kırklareli, Thrace), where the sugar factory was located, had been transformed from "the land of malaria" into "a masterpiece of the new and modern Turkey with its mechanized and civilized appearance" (Yayman, 1936: 10). In the next section, I explore the historical relationship between industrial humanities and regional identities during the early Republican era.

The labor hinterland of Alpullu Sugar Factory

Alpullu deserves special attention within these regional narratives since Alpullu Sugar Factory, founded in 1927 to produce sugar, molasses, and alcohol, was Turkey's first. The factory employed German and Hungarian engineers until 1936, when high-ranking positions were assigned to Turkish citizens. The factory settlement was built like an industrial housing complex, with multiple social and sports facilities, including a swimming pool and golf course, workers and management housing, and a 20-bed hospital with a surgery. There was a primary school and later a high school, named Sugar School (*Şeker Okulu*), and an Art School for training workers and engineers for the sugar industry. Thus, Alpullu Sugar Factory was like a town, designed to develop a community from within, which was further planned to serve as a center for commodity and service flows.

During a fieldtrip to Alpullu in 2012, I interviewed six former workers at the factory.[14] The main emphasis was on the high culture and modernity that the factory brought to the town. For example, while talking about the sports facilities located at the factory settlement, one of the participants vehemently claimed, "Alpullu was like a European town . . . an Olympic village." According to another interviewee, the company that owned the factory and its settlement[15] established a market in 1935, known as the factory's commissary store, close to the railway station, to sell groceries and dairy produce (especially butter, eggs, and milk). The factory also owned a small dairy farm to provide the workers' meals. Due to circulation of consumer products, which were not found anywhere else at that time, "people used to call Alpullu, 'the center' 30 to 40 years ago."

The factory, as a place was not only a vehicle to eradicate the Ottoman past but was also depicted as a space where the dream of the nation came true. The fact that Turkish engineers worked together in such factories to develop Turkey's economy framed the factory-place as a microcosm of the nation. The representations of the Alpullu factory, which blended civilization and Turkification through the lens of Turkish modernity, were especially powerful in the early Republican era. Ratip Tahir's 1934 painting (Figure 4.1), which used to hang in the local school, symbolized this amalgamation of the Republican elite gaze at the factory. The painting portrays Alpullu's *Mimar Sinan Köprüsü* [Mimar Sinan Bridge], constructed in the sixteenth century, but still standing in the twentieth century, as a symbol highlighting the robustness of the factory – which is framed by the bridge arch, with both presented as cornerstones of Turkish civilization and the discourse of Turkish migration. In the 1960s, the painting was reproduced in a photograph (Figure 4.2) taken by the factory's former photographer, who was one of the interviewees during my visit to Alpullu. By highlighting the bridge with the factory, the painting and photo reiterate the idea that industrialization was a tool to fight imperial disorderly nature.

The new, independent nature of the Turkish Republic was narrated through the discursive rationalization of nature. Factory-places were regarded as former areas of malaria and marsh, associated with the neglect that characterized Ottoman rule. These marshes were re-narrated as places of production in the Republican era. Factories and modern homes were instrumental tactics for producing a rational environment and nationalist ideology. The war against nature was narrated along with returning to the origins of Turkishness, where the idea of civilization was supposedly embedded. Thus, taming nature became a fight against the Ottoman rule and the colonial powers.[16]

The early Republican policy of urbanization through industrial settlements was shaped by an urban-rural dichotomy. Its emphasis on the need for modernizing rural areas illustrates a key Kemalist ideology promoted during this period. In

Figure 4.1 Oil painting of the Alpullu Sugar Factory

Mimar Sinan Köprüsünden Alpullu Şeker Fabrikası'nın görünüşü (1

Figure 4.2 Photo of the Alpullu Sugar Factory

Trakya, rural areas and rurality are despised and understood as the result of Ottoman neglect whereas rural development, urbanization, and/or modernization of
villages are presented as the ultimate target for creating a westernized Turkish
nation. It is easy, however, to see the contradiction within this ideology. For
example, although western culture is embedded in Turkish history, it is still something to be targeted. In that sense, the nationalism articulated in the journal ter-
ritorialized the nation as in need of modernization to become urban by promoting
the eradication of the Ottoman past.

During the early Republican period, the statist policies shaping industrialization and urbanization were intended to form the organic unity of the nation by
consolidating territorial control, while further initiating governance and disciplining of hierarchies across ethnic, class, and urban–rural divides (İnsel, 1996;
Yalman, 2002). Territorial control was not only limited by the ways the factory
was represented as part of a modern nation-state spatial practice in *Trakya*. In
Alpullu, too, factory design and operations aimed at territorializing class differences. The factory's gated housing area, which could not be entered without
management permission, was named "the colony" (*koloni*). Here, the term "colony" perhaps associated the settlement with an ideal national community, and
was especially targeted to influence people living in Alpullu or neighboring villages. In my casual conversations, usually with people outside Alpullu, Alpullu's
social and sports facilities were often remembered for their restrictions and/or
people's furtive use of them. Thus, the boundaries of the factory settlement must

have been instrumental in distinguishing classes of people from each other. Similarly, hierarchy was evident in the separation of cinema audiences by time, as one interviewee recalled: "Wednesday screenings were for the engineers, working foreman, people that they [we] called the elites (*protokol*)." At the other screenings, even if they were accessible to the people of Alpullu, as another participant pointed out, "Still the ones without proper dress would not be allowed." With their exclusive sports and social facilities, and their class and employment hierarchies, factory spaces were also instrumental cultural projects to develop Turkey's modern national culture (Peri, 2002; Karakaya, 2010; Keskinok, 2010). Such factory-places further aimed to introduce class-based modern living, as specifically observed in the factory settlement and Alpullu town.

Within this overall nationalist context, commercial and industrial places, modern homes, urban networks, and parks became important symbolic representations of the national landscape. Construction of these places came with their glorification and even fetishization, at least representationally. They were also seen as places of the Turkish nation. In Thrace, the emphasis on migration was territorially worked by the Turkish state to transform former Ottoman lands into Turkish state territory in which the factory settlement, modern home, modern village, and/or village rehabilitation of the villages found national materiality.

Conclusion

The Republican era was a turning point in territorializing the nation while re-designating socio-spatial proximities across different ethnic and class groups. This process founded the new Turkish nation and state based on a re-narration of history (the origins) and geographical limits. Here, geography means more than border making with other countries for legitimizing territorial sovereignty as it also includes boundary making within the country itself. I explored the social and discursive constructions of Turkish territory and territoriality, and the ways through which these constructions were co-constituted across the historical narratives and places of the "imagined" Turkish nation. The emphasis on migration, strengthened by the power of the settlement law were important state tactics to transform former Ottoman lands into Turkish state territory within which the factory settlement, modern home, modern villages, and/or rehabilitation of villages found national materiality. These places all symbolized the Turkish takeover of Anatolia as factory settlements and modern villages were among the new venues of Turkish modernity. These places were widely mobilized by the Inspectorship in Thrace. The contextual emphasis on constructing the new territory through economic spaces, such as factories and ideal villages, shaped ideas of nature. The state elites and their representatives in the Inspectorship used the taming of nature to represent a rupture from the Ottoman Empire.

The socio-spatial trajectories of territory-building engage with the meanings and materialities of nature in all of its meanings (Williams, 1983 [1976]). Therefore, they are also important for the environmental history. This kind of an environmental history account which I focus here is more interested in understanding

historical trajectory as a means to explore territorial, national, and public policies. Future studies can consider whether discourses of modernization and taming
nature in Turkey's case persist alongside ideas and representations of the Turkish
nation and/or explore how these ideas have became diversified and were disseminated within the vast fields of environmental criticism (i.e. environmental
movements).

Notes

1 The English and Turkish words for nature have something in common since the Turkish word for nature, *doğa*, comes from the root "to be born". It also means the essential
   character (*mizac, huy*) of something, which is foundational and cannot be changed.
   It further encompasses the entirety of human and non-human beings in the material
   world. Lastly, it is the force that governs the material world. For more see Harris (2014).
2 See also Ersoy (2017).
3 Koçak (2003) stresses that the era coincided with the Armenian Genocide, by which
   he implies that the inspectorships indirectly contributed to it.
4 Moreover, public inspectorships were allegedly adapted to suppress the Kurdish conflict in the 1990s in the form of the governorship of the State of Emergency of the
   Region (*Olağanüstü Hal Bölge Valiliği*) between 1987 and 2002 (Koçak, 2003).
5 The first Public Inspectorship was responsible for securing and modernizing Elazığ,
   Urfa, Bitlis, Hakkari, Diyarbakır, Siirt, Mardin, and Van. The second Public Inspectorship was established in Thrace, covering Edirne, Tekirdağ, Kırklareli, and Çanakkale. The third, established in 1935, governed security and regional development of
   several north-eastern cities (Erzurum, Erzincan, Trabzon, Ağrı, Gümüşhane, Kars, and
   Artvin). The fourth was established solely for Dersim (Tunceli) in 1935, following an
   aerial bombardment to suppress a regional revolt. Finally, the fifth Public Inspectorship, established in 1947, covered the southern provinces (Hatay, Adana, İçel, Maraş,
   and Antep) (Koçak, 2003). The locations of these five public inspectorships make it
   clear that they were focused on governing and securing the politically and socially
   restless border regions of the newly established Turkish Republic.
6 For example, Kurds were frequently described as Turks who had gradually forgotten
   their "true" ethnic identity. Non-Turkish identities were paradoxically both recognized and disavowed. Otherwise, there would have been no need for such emphasis on
   Turkish identity as the source of national belonging.
7 For an analysis of "territorial nationalism" and the mental and physical meanings of
   the nation (*vatan*) constructed by geography books from 1912 to the early 1930s, see
   Özkan (2014).
8 From the early eighteenth century onwards, the imperial territories covered three continents with ongoing conflicts and clashes that forced Ottoman bureaucrats backed by
   local gentry to reconfigure administrative tactics to contain regional and/or local riots
   (Salzmann, 2004).
9 *Vatan* stands for one's place of birth in Arabic but its meaning was changed to refer
   to national territory and/or homeland in different nationalisms in the Middle East
   (Najmabadi, 1997; Özkan, 2014).
10 Among these reforms, establishing a national education system and adopting the
   Latin alphabet were particularly influential in nation-building. For more, see Ahmad
   (1993), Özkan (2014), and Altınay (2004).
11 Istanbul, Bursa, and Tekirdağ were the leading cities for settling Muslim-Balkan immigrants (Geray, 1970; Sosyal Planlama Başkanlığı, 1990).
12 This zone also encompassed places where Armenians had lived before the genocide in
   1915.

13 It is important to note that the Dersim Massacre took place in 1937 when the Turkish military put down a revolt against the Settlement Law. In 1937, the government enforced the law in Dersim by relocating local people to ethnically homogenize the region (Hür, 2011).

14 The semi-structured and exploratory interviews were approximately about an hour in length, focusing on the past memories of the Alpullu factory and failed attempts of its privatization. Currently, Alpullu is not privatized thanks to a municipal campaign, backed up by community support.

15 According to my interviews at the local municipality, almost two thirds of the town's land is owned by the company that also owns the factory itself.

16 Izmir expo site was constructed at the center of Izmir in 1936. It was a site, burnt by the Greek army while retreating from Anatolia before the Republic was established. For the meaning of its location, the Republican elites saw the expo site as a symbol of an emphasis on industrialization and rationalization, as a means to struggle against colonial powers of the West (see also İnan, 2005: 78).

Bibliography

Adanır, F. (1996) [1979]. *Makedonya Sorunu*, trans. İhsan Catay. Istanbul: Tarih Vakfı Yurt Yayınları.

Agnew, J. (1994). The Territorial Trap: The Geographical Assumptions in International Relations Theory. *Review of International Political Economy*, 1(1), pp. 53–80.

Ahmad, F. (1993). *The Making of Modern Turkey*. London: Routledge.

———. (2008). Politics and Political Parties in Republican Turkey. In R. Kasaba, ed., *The Cambridge History of Turkey, Turkey in the Modern World*, 1st ed. Cambridge and New York: Cambridge University Press, pp. 226–265.

Akşit, E. E. (2005). *Kızların Sessizliği*. Istanbul: İletişim Yayınları.

———. (2010). Politics of Decay and Spatial Resistance. *Social & Cultural Geography*, 11(4), pp. 343–357.

Altınay, A. (2004). *The Myth of the Military-nation: Militarism, Gender, and Education in Turkey*. New York: Palgrave Macmillan.

Anderson, B. (1991). *Imagined Communities: Reflections on the Origin and Spread of Nationalism*. London: Verso.

Anonymous. (1936a). Baykuşlar ve Padişahlar. *Trakya*, 1, p. 22.

———. (1936b). İmece. *Trakya*, 1(2), pp. 4–5.

———. (1936c). Özlü Sözler. *Trakya*, 1(4), p. 18.

———. (1937). Özlü Sözler. *Trakya*, 1(6), p. 21.

Aslan, Ş., Arpacı, M., Gürpınar, Ö., and Yardımcı, S. (2014). *Türkiye'nin Etnik Coğrafyası*. Istanbul: Mimar Sinan Güzel Sanatlar Üniversitesi.

Baydar Nalbantoğlu, G. (1997). Silent Interruptions: Urban Encounters with Rural Turkey. In S. Bozdoğan and R. Kasaba, eds, *Rethinking Modernity and National Identity in Turkey*, 1st ed. Seattle and London: University of Washington Press, pp. 192–211.

Bozdoğan, S. (2001). *Modernism and Nation Building, Turkish Architectural Culture in the Early Republic*. Seattle and London: University of Washington Press.

Brenner, N. (2004). Urban Governance and the Production of New State Spaces in Western Europe, 1960–2000. *Review of International Political Economy*, 11(3), pp. 447–488.

Brenner, N. and Elden, S. (2009). Henri Lefebvre on State, Space and Territory. *International Political Sociology*, 3, pp. 353–377.

Burgaç, M. (2013). *Türkiye Umumi Müfettişliklerin Kurulması ve Trakya Umumi Müfettişliği*. Ankara: Atatürk Araştırma Merkezi Yayınları.

Cagaptay, S. (2003). Citizenship Policies in Interwar Turkey. *Nations and Nationalism*, 9(4), pp. 601–619.

———. (2006). *Islam, Secularism, and Nationalism in Modern Turkey: Who Is a Turk?* New York: Routledge.

Castree, N. and Braun, B., eds. (2001). *Social Nature: Theory, Practice and Politics*, 1st ed. Oxford; Malden: Blackwell Publishing.

De Carvalho, B. (2016). The Making of the Political Subject: Subjects and Territory in the Formation of the State. *Theory and Society*, 45, pp. 57–88.

Dirlik, K. (1936a). El Birliği: Toplu Çalışmak en verimli kudrettir. *Trakya*, 1(2), pp. 1–2.

———. (1936b). Köy Kalkınması Hakkında Düşünceler. *Trakya*, 1(3), p. 12.

Elden, S. (2007). Governmentality, Calculation, Territory. *Environment and Planning D: Society and Space*, 25, pp. 562–580.

———. (2010). Land, Terrain, Territory. *Progress in Human Geography*, 34(6), pp. 799–817.

———. (2013). *The Birth of Territory*. Chicago: The University of Chicago Press.

Ersoy, M. (2017). *Osmanlıdan Günümüze İmar ve Yasalar*. Istanbul: Ninova Yayınları.

Evered, K. T. and Evered, E. Ö. (2012). State, Peasant, Mosquito: The Biopolitics of Public Health Education and Malaria in Early Republican Era. *Political Geography*, 31, pp. 311–323.

Fairclough, N. (1992). *Discourse and Social Change*. Cambridge: Polity Press.

———. (2003). *Analysing Discourse: Textual Analysis for Social Research*. London; New York: Routledge.

Foucault, M. (1974). *The Archaeology of Knowledge*. London: Tavistock.

———. (1989). *The Archaeology of Knowledge*. London: Routledge.

———. (1991). *Foucault Effect: Studies in Governmentality, With Two Lectures by and an Interview with Michel Foucault*. P. Miller, G. Burchell, and C. Gordon, eds. London: Harvester Wheatsheaf Press.

Geray, C. (1970). Türkiye'de Göçmen hareketleri ve Göçmenlerin Yerleştirilmesi. *Amme İdaresi Dergisi*, 3(4), pp. 8–36.

Hakli, J. (2008). Regions, Nteworks and Fluidity in the Finnish Nation-State. *National Identities*, 10, pp. 5–22.

Hall, S. (2001a). Foucault: Power, Knowledge and Discourse. In M. Wetherell, S. Taylor, and S. Yates, eds, *Discourse, Theory and Practice*, 1st ed. London: Sage Publications, pp. 72–81.

———. (2001b). Foucault: Power, Knowledge and Discourse. In M. Wetherell, S. J. Yates, and S. Taylor, eds, *Discourse Theory and Practice: A Reader*, 1st ed. Thousand Oaks, CA: Sage, pp. 72–82.

Harris, L. M. (2014). Imaginative Geographies of Green: Difference, Postcoloniality, and Affect in Environmental Narratives in Contemporary Turkey. *Association of American Geographers. Annals of the Association of American Geographers*, 104(4), pp. 801–815.

Hür, A. (2011). 1934 İskan Kanunu ve Kürtler. *Taraf*. [online] Available at: www.taraf. com.tr/ayse-hur/makale-1934-iskan-kanunu-ve-kurtler.htm [accessed on November 6, 2013].

İnan, A. (2005). *Prof. Dr. Afet İnan*. Istanbul: Remzi Kitabevi.

İnsel, A. (1996). *Düzen ve Kalkınma Kıskacında Türkiye, Kalkınma Sürecinde Devletin Rolü*. Istanbul: Ayrıntı Yayınları.

Jones, R. (2007). *People/State/Territories*. Oxford; Malden, MA: Blackwell.

Jongerden, J. (2009). Crafting Space, Making People: The Spatial Design of Nation in Modern Turkey. *European Journal of Turkish Studies*, 10, pp. 1–22.

Kadıoğlu, A. (1998). *Cumhuriyet İradesi Demokrasi Muhakemesi*. Istanbul: Metis Yayınları.

Kale, B. (2014). Transforming an Empire: The Ottoman Empire's Immigration and Settlement Policies in the Nineteenth and Early Twentieth Centuries. *Middle Eastern Studies*, 50(2), pp. 252–271.

Kandemir, S. (1936). Köy Nedir?. *Trakya*, 1, pp. 4–5.

———. (1937). Elbirliği. *Trakya*, 1(6), pp. 15–16.

Karakaya, E. (2010). Construction of the Republic in City Space: From Political Ideal to Urban Planning. M.A. Ankara: Middle East Technical University.

Karaömerlioğlu, A. (1999). The People's Houses and the Cult of the Peasant in Turkey. In S. Kedourie, ed., *Turkey Before and After Atatürk*, 1st ed. London; Portland: Frank Cass Publishers, pp. 67–91.

Karpat, K. (2012). *Balkanlar'da Osmanlı Mirası ve Milliyetçilik*, trans. Recep Boztemur, 2nd ed. Istanbul: Timaş Yayınları.

Keskinok, Ç. H. (2010). Urban Planning Experience of Turkey in the 1930s. *METU Journal of the Faculty of Architecture*, 27(2), pp. 173–188.

Kirişçi, K. (2000). Disaggregating Turkish Citizenship and Immigration Practices. *Middle Eastern Studies*, 36(3), pp. 1–22.

Koçak, C. (2003). *Umumi Müfettişlikler (1927–1952)*. Istanbul: İletişim Yayınları.

Nagel, J. (1998). Masculunity and Nationalism: Gender and Sexuality in the Making of Nations. *Ethnic and Racial Studies*, 21(2), pp. 242–269.

Najmabadi, A. (1997). The Erotic Vatan [Homeland] as Beloved and Mother: To Love, to Possess, and To Protect. *Comparative Studies in Society and History*, 39(3), pp. 442–467.

Öney, S. (1936). Köy Kısaca Türkiye. *Trakya*, 5, pp. 15–16.

———. (1937). Kalkınma. *Trakya*, 6, p. 7.

Özkan, B. (2014). Making a National Vatan in Turkey: Geography Education in the Late Ottoman and Early Republican Periods. *Middle Eastern Studies*, 50(3), pp. 457–481.

Paasi, A. (2003). Territory. In J. Agnew, K. Mitchell, and G. Toal, eds, *A Companion to Political Geography*, 1st ed. Oxford: Blackwell, pp. 109–122.

Painter, J. (2010). Rethinking Territory. *Antipode*, 42(5), pp. 1090–1118.

Parla, A. (2005). Terms of Belonging: Turkish Immigrants from Bulgaria in the Imagined Homeland. Ph.D. New York: New York University.

Peri, B. (2002). Building the Modern Environment in Early Republican Turkey: Sümerbank Kayseri and Nazilli Factory Settlements. M.A. Ankara: Middle East Technical University.

Radkau, J. (2008). *Nature and Power: A Global History of the Environment*. New York: Cambridge University Press.

Raffestin, C. and Butler, S. A. (2012). Space, Territory, and Territoriality. *Environment and Planning D: Society and Space*, 30, pp. 121–141.

Salzmann, A. (2004). *Tocqueville in the Ottoman Empire*. Leiden; Boston: Brill.

Sayar, M. Z. (1952). Hatalı Bir İskan İşi. *Arkitekt*, 5 (249-250-251-252), pp. 101–102.

Şengül, T. (2003). On the Trajectory of Urbanisation in Turkey, an Attempt at Periodisation. *IDPR*, 25(2), pp. 153–168.

———. (2009). *Kentsel Çelişki ve Siyaset, Kapitalist Kentleşme Süreçlerinin Eleştirisi*. Ankara: İmge Yayınevi.

Şimşek, Sefa. (2002). *Bir İdeolojik Seferberlik Deneyimi Halkevleri 1932-1951*. Istanbul: Boğaziçi Üniversitesi Yayınevi.

Sirman, N. (2007). Kadınların Milliyeti, Milliyetçilik. In T. Bora and M. Gültekingil, *Modern Türkiye'de Siyasi Düşünce, vol. 4*. Istanbul: İletişim Yayınları.

Sosyal Planlama Başkanlığı. (1990). *Bulgaristan'dan Türk Göçleri*. Ankara: T.C. Başbakanlık Devlet Planlama Teşkilatı Sosyal Planlama Başkanlığı.

70   *Eda Acara*

Tekeli, İ. (2013). *Modernizm, Modernite ve Türkiye'nin Kent Planlama Tarihi*. Istanbul: Tarih Vakfı Yurt Yayınları.

Williams, R. (1983) [1976]. *Keywords, A Vocabulary of Culture and Society*. New York: Oxford University Press.

Yalman, G. (2002). Tarihsel Bir Perspektiften Türkiye'de Devlet ve Burjuvazi: Rölativist Bir Paradigma Mı Hegemonya Stratejisi Mi? *Praksis*, 5, pp. 7–23.

Yayman, K. O. (1936). Yürüyen Devrim. *Trakya*, 1(4), pp. 9–11.

Part two

State and capital on the edge

5 Security, dispossession, and industrial meat production in Turkey[1]

Sezai Ozan Zeybek

Livestock production in Turkey has drastically changed over the last couple of decades from village type, free range production to industrial mass production. Animals now live and are being killed in a different bio-spatial regime. These changes, which I will analyze here, have two dimensions: The first concerns the set-up through which markets, states, and supra-national institutions intervene in ecological networks and production regimes in order to increase and accelerate capital accumulation. Their interventions are entangled with normative claims about efficiency, growth, and profitability. The second dimension regards the Turkish state's security concerns over the Kurdish conflict, which resulted in wiping off/displacing 'uncontrolled' human and animal populations for the sake of security in the 1990s and replacing them with easily controllable, closed-off, industrial facilities. Therein, two distinct agendas i.e., 'secure zones' and 'industrial food chains', are being pulled into one operational logic.[2] Those humans and non-humans who do not fall in line with the demands of the latest production regime are either discarded or criminalized, or often both. As a result, animal breeds, technological infrastructure, and the configuration of power have been radically transformed. New categories and a new repertoire have emerged in terms of population, life, and efficiency.

I claim that these changes are not the result of conventional market mechanisms. On the contrary, non-market devices are pushed ahead in favor of the more powerful actors in the field: "The capitalist enterprise is indeed a means of employing nonmarket arrangements to produce goods and services for the market" (Davis and Monk, 2008: 26). It has a strong tendency towards monopolization and operates through a set of stratagems that violate any self-attributed understanding of 'free market'. In reference to Fernand Braudel, Manuel de Landa defines anti-markets as a set of arrangements where every aspect is routinized and firmly controlled, including the labor force, which stands in stark contrast to autocatalytic developments, open competition, and non-centralized creativity. He argues that capitalism is founded not on free markets as such, but on their restriction, and even periodical annulment (De Landa, 2000).

In this respect, I wish to highlight the intersecting trajectories of deepening capitalization and military violence in Southeastern Turkey. I show how dispossession takes place in a particular military setting and how bio-political regimes align with not-free-market mechanisms. To that purpose, I look at how animals are subjected to regulative, normative policies of states (more than one state) and corporations. As Michel Foucault has taught us, the exercise of power cannot be limited to a solely human affair. It is rather practiced within the nexus of apparatuses, including bodies of both humans and non-humans and technologies (Foucault, 1979, 1988; see also Deleuze, 2006).

Below, I begin with a discussion of the new ecosystem that emerge around a particular legal framework and a new mode of capitalization, while putting a number of normative claims under scrutiny. Then, I examine the spatial arrangements in the above-mentioned conflict zone and the subsequent demographic transformation (of both humans and non-humans) that has happened for the past 40 years. My main argument is the following: Controlling animal populations has been key for controlling human populations from a state's point of view. Increasing rates of capital accumulation thus depend on disciplining, monitoring, and supervising bodies (of animals) and rendering them more 'efficient' and more 'secure' (Foucault, 1991: 210–212).

Livestock productivity – a critique

At a symposium held in 1987, the well-known Turkish economist Gülten Kazgan gave a *tour d'horizon* about what Turkish agriculture should look like in the year of 2000 (Kazgan, 1988). She argued that emerging bio-technologies – that is, the modification of genes – would soon present a break-through in agriculture, an innovation similar to the Green Revolution – a new mode of mechanized agricultural production that involves the use of pesticides and fertilizers. She claimed that the Green Revolution had enabled developed countries to become self-sufficient in terms of food, while turning developing states dependent on them. The reason she put forward was the dumping of agricultural over-production from richer onto poorer countries, creating a supply with which local producers could not compete (Kazgan, 1988: 260; for a thorough analysis of the same process, see McMichael, 2000). Similarly, she argued, upcoming innovations were to amplify existing inequalities between importer and exporter countries and purge inefficient segments in the production chain. Therefore it was vital for Turkey to get a foothold in the race and open up domestic markets to foreign competition (1988: 259).

It is not quite clear how opening markets would encounter the threat she mentioned, but apparently, her proposed game plan was not to shy away from adopting new capital-intensive technologies in agriculture. After the financial crisis at the end of the 1990s and due to the ensuing austerity policies imposed by the International Monetary Fund (IMF), Turkish agriculture has indeed become a more capital-intensive sector run by higher-value inputs. By introducing hybrid seeds and new breeds of livestock, the entry barrier to the market has been raised. Put more simply, the more the seeds/animals/animal feed has increased in price, the harder it has become for small farmers to stay in production. In order to sustain their competitiveness, farmers had to invest more in bio-technologies.

Although developmental approaches would applaud such investments, I would argue that higher prices also serve to reduce competition and enhance capitalization by the few, as will be illustrated below.

However, the real issue with Kazgan's call for competition was her unquestioning acceptance of several key assumptions of the economic rationale, such as efficiency, comparative advantage, and growth.[3] As with security, these concepts should above all be tackled as political claim-making devices, not as neutral reflections of reality.[4] The way one calculates and compares 'growth' or 'efficiency' is not as straightforward as it may seem at first. Constructing these terms needs a series of strategic omissions and the imposition of contestable priorities.

Let me give an example of possible versions of calculating efficiency in livestock. Livestock breeds that are currently subsidized by the Turkish state – with the two prominent ones being Holstein and Jersey in milk races – yield annually 5,900 liters of milk. In comparison, indigenous breeds can only give 1,800–1,900 liters (Aysu, 2016). With only these two figures at hand, it makes sense for policymakers to promote imported 'high-yield' breeds as opposed to indigenous ones; in fact, this has been a decade-long policy of Turkey. But many factors are being rendered invisible in the above comparison, because each living being is connected to its environment in intricate ways irreducible to milk production only. To start with, a Yerlikara cow, an indigenous breed, can calve 8–9 times during her life span, whereas a Holstein cow can do so only 2–3 times.[5] The productivity of a Holstein cow often decreases by 40 percent after a few years. In order to maintain her high-level productivity, she is given protein-based nutrition instead of fodder, which is supplied by a larger commodity chain and therefore more expensive for the farmer.[6] The total area dedicated to this type of feed-crop production amounts to one-third of total arable land globally (Steinfeld et al., 2006: xxii). Furthermore, to sustain this so-called high productivity, animals are supplemented with growth hormones, particularly Bovine Somatotropin (BST), which is banned in the European Union (EU), but still in use in Turkey. Cows that are subjected to these bio-technological improvements now yield on average three times more milk than their predecessors in the 1950s (E. Marcus, 2005: 10–11).

By contrast, 'inefficient' breeds can wander around and feed themselves on grass growing on small patches. Their manure can turn into fertilizer, or energy, and support autonomous production cycles. A Holstein cow's manure, however, would not be fit for such purposes, as it contains toxic chemicals (such as antibiotics) and would kill living organisms if dumped onto the soil unprocessed (Aysu, 2016).[7] Massive volumes of manure coming from concentrated zones are beyond the composting capacity of the soil – or rather of worms, mushrooms, and bacteria. It first has to be processed by humans, creating an additional expense, or, as in Turkey, it is simply moved to someone else's backyard. Concentrated manure dumped in this manner would pollute and poison rivers and underground water (Steinfeld et al., 2006; for a more detailed discussion on manure, see: 136–149).

There are further ramifications of switching from one production scheme to another. But even this single example illustrates how efficiency is calculated with a very limited set of parameters. Efficiency, as such, favors certain actors over others. Particular criteria, such as volume or quantity, are seen as the most important measure and are used to discharge so-called inefficient processes, people, and breeds. Social justice, health, or resilience do not necessarily constitute the core values of the capital-intensive industrial production. Within its perverse logic, even the deterioration of some public goods (such as health) may count as another opportunity for profit accumulation. The weakened immune systems of animals that are forced to live in feedlots augment the profits of medical corporations and add to their growth figures.

In sum, neither 'growth' nor its burdens are distributed evenly. The current techno-political apparatus favors bigger investors who can externalize a significant share of their costs. Below, I will follow the trajectory of the recent transformations in Turkey which endorsed the increase in costs in the agricultural sector and, as a result, the dispossession of a significant number of producers.[8]

New avenues of capitalization in agriculture

It is hard to assign a scale to the agricultural transformations in Turkey, or to determine a year zero for the shifts that occurred. However, basic economic indicators give an overall idea. The last wave of de-ruralization in Turkey (as the flipside of urbanization) took place very rapidly and sharply, reducing the weight of agricultural exports to 3.7 percent as of 2015, a fraction of what it used to be in the 1970s, when it was about 70 percent. From the 1980s onwards, upon the advice of the IMF, Turkey adopted an export-led growth policy, and part of the deal was to expose domestic production to external competition. As a result, between 1980 and 1990, exports in agriculture increased by 24 percent; yet, in the same period the surge of imports reached 1,440 percent (tuik.gov.tr, 2015a). Today, agriculture in Turkey is carried out under a new spatial regime, which is dependent on larger commodity chains from seeds to basic consumer items.

Arguably, the most recent turning point of these changes happened right after the financial crisis that Turkey experienced at the end of the 1990s. To overcome the crisis, the government responded to IMF's demands (1999). Total amount of subsidies were reduced by almost half in the subsequent years and replaced by direct income provisions, as assured in the Letter of Intent (Oyan, 2013: 123; Zeybek, 2016). The pretext for doing so was the inefficiency of the existing subsidy programs. Prices in markets, it was told, were affected negatively by the state's interventions, and most importantly, existing schemes favored mostly big farmers over small ones (Hazine Müsteşarlığı, 2000).

It is often held against the IMF that these policy recommendations were not specific to Turkey, but rather endorsed in different countries as ready-made recipes. In fact, these programs were not applied ubiquitously, but limited to particular countries. In contrast to developing countries, a variety of support schemes were effectively in use in the EU. While the weight of direct income support (as foreseen by the IMF) was elevated to 83 percent in Turkey in the post-crisis years, it never surpassed 30 percent in the EU region (Oyan, 2013: 121–122). Instead, input, credit, price, and infrastructure subsidies were used extensively without worrying about 'distorted markets'.

IMF policies systematically diverted money from public expenditures to mostly private debtors and facilitated a new spatial organization in production. Direct income was given to land owners based on fixed rates – that is, proportional to the land size. This meant that low-volume, and possibly poly-culture production in small patches (such as fruit gardens, greenhouses, and vineyards), could benefit less from this scheme than high-volume mono-culture production conducted on huge tracts of lands. A report published by the Ministry of Agriculture and Forestry in 2006 stated that 51 percent of all the money went to only 17 percent of farmers who owned more than 100 hectares of land. In other words, less than half of the budget had to be shared by 83 percent of farmers. Another report by the World Bank in 2004 showed that the new scheme could compensate for only half of the net income loss of farmers (both reports from Oyan, 2013: 123). Thus, direct income was not really designed to give support to small farmers or to support agriculture in general, but used to dampen possible social reactions against it. In other words, it was a social benefit scheme. In the end, the government could not sustain this controversial change and had to shelve it after no more than five years.[9]

Turkey also pledged in the Letter of Intent to roll back the state's direct involvement in agricultural production. This meant privatization of the public assets and institutions. Slaughterhouses, feed complexes, and fertilizer factories, cold storage units were either shut down or sold (esk.gov.tr, 2018). Small-scale farmers have become more vulnerable to market forces. In livestock production, the ministry decided to allocate all its budget to farms with more than 40 animals (Serhat Kalkınma Ajansı, 2011: 69). Farms with 10 to 50 heads of cattle were assigned to the Regional Development Agencies (*Bölgesel Kalkınma Ajansları*), founded upon the initiative of the EU. At that time (2010), however, 76 percent of all milk-producing cattle farms in Turkey had less than 10 animals (TMMOB Ziraat Mühendisleri Odası, 2015). They were not eligible to receive support from either ministry or development agency. In a way, the re-structuring of subsidies for livestock farming was devised to phase out small farmers and to channel all the funding to larger producers (Ibid.).

The outcome was not surprising. In 2002, there were around 4,000 farms with more than 50 heads of cattle. Over 10 years, this figure increased by 550 percent, to 24,000 (Mehdi Eker, then Minister of Agriculture, quoted in Türkiye'de 50 Baş ve Üzeri Hayvan Varlığı Olan 24 Bin İşletme Var, 2012). Farmers were expected either to borrow money and invest in buying more animals, or to quit production. Many could not continue farming. In this period, approximately 2.5 million farmers (not just livestock farmers) gave up agricultural production (Oyan, 2013: 127). Over 15 years, cultivated lands decreased by the size of Belgium, as 4 million hectares of land were abandoned (Ali Ekber Yıldırım, journalist/expert on agriculture, quoted in Karadeniz, 2015). What these uncultivated lands may turn into is still an open question. Banks and corporations started buying these lands allegedly for extracting minerals or for future construction projects (roads, new settlements, dams) (Döner, 2016).

As I have argued above, this transformation happened not as a consequence of the free markets, but because of the incessant interventions of a number of institutions. Certificates, subsidies, interest rates, counseling firms, development agencies, and the like have offered new avenues for conducting agriculture and restricted others. Notions such as health, hygiene, efficiency, and competitiveness played a key role in legitimizing these transformations and, hence, the concentration of capital accumulation in fewer hands. Animals were enclosed and subjected to efficiency-enhancing bio-chemical processes as well as to the disciplines of the industrial complex. In parallel, old forms of farming have become more and more insecure and obsolete. Farmers have had to abandon their lands due to economic pressures and military violence.

Until now, I have focused on the former. In the following section, I will mainly look at the 1990s, a period in which forced evacuations and the military conflict in Turkey's Kurdish-dominated region led to a drastic change in the production regime. I argue that the bio-political control of populations, both human and animal, has been facilitated by violence and enhanced capitalization on the one hand and dispossession on the other.

Killing animals for the sake of security

The 1990s were a breaking point in the Kurdish-dominated region of Turkey.[10] The long-standing armed conflict between the Kurdistan Workers' Party (PKK; *Partiya Karkerên Kurdistanê*), the Kurdish guerrilla movement, and the Turkish army took a new turn when the Turkish state decided to evacuate and destroy villages as a counter-insurgency measure (Jongerden, 2010). In the meantime, the army put pressure on villagers to join the pro-state militia in the fight against the PKK. This was meant as a policy to create divisions among the Kurds as well as to determine the political allegiance of each village. Those who refused to join the fight against the PKK had to leave. In the end, according to some reports, 4 to 5 million people were affected (Barut, 2001). Government figures, however, reported the number of displaced persons to be as low as 384,793, with only 3,215 evacuated settlements (Jongerden, 2010: 79). The Institute of Population Studies at Hacettepe University is considered to have published the most reliable report on this matter. Accordingly, 953,680 to 1,201,200 individuals were forced to leave their settlements for reasons of 'security' between 1986 and 2005. The affected area mentioned in the report covered 14 districts in Southeastern Turkey. Of the affected population, 80 percent were from rural areas (Çağlayan et al., 2011: 27). A large part of the internal displacement took place within the region, from villages and hamlets to regional cities. For example, the population of Hakkari deep in the southeastern corner of Turkey doubled between 1990 and 2000. In the same manner, relatively small urban settlements such as Kızıltepe, Şırnak, and Batman entered the new millennium as over-crowded urban centers. In Turkey, a census is supposed to be conducted every five years. Remarkably, no population census was held in 1995, in the heat of forced evacuations. Then, in 1997, when a census was finally carried out, it was immediately declared null and void on the grounds of technical mistakes (Yılmaz, 1999). Therefore, we cannot really establish the true extent of the displacement.

There are many facets of this period that need to be studied as of yet. In order to understand the every so often spiraling violence and rapid shifts within Turkish society, one has to analyze and account for these experiences in relation to class, gender, religious affiliations, ethnic ties, and so on (for already existing, important studies, see Çağlayan et al., 2011; Baysal, 2014; Yağız et al., 2012; Mutlu, 2015). However, my point here is that the violence has not merely been directed at humans, but at ecosystems. By this term, I mean the larger ecological networks that consist of both humans and non-humans. To deter people from returning to their lands, trees were burned down, gardens destroyed, forests set on fire, and animals massacred.

> [PKK] guerrillas hit the military post in Celeka on the slope adjacent to our village [near Siirt]. Soldiers were killed, I don't know how many. Those who survived asked for back-up. They came with helicopters. Most villagers fled. When the soldiers arrived, they killed the eight people who had stayed. Then they killed all of our animals, thousands of them. To prevent a pandemic, they threw the dead bodies into the river. The pile [of bodies] was so big that the river was blocked.
>
> (Cemile from Siirt, quoted in Yağız et al., 2012: 165–166)

These aspects are still understudied in their full scale, even though they do come up in reports or emerge in personal narratives as 'secondary anecdotes'. Husbandry had been a key economic activity in the region. It was traditionally conducted not in enclosed or industrial, but in open spaces, through the herding of goats and sheep especially. However, due to the land mines placed in the prairies, herding animals became a life-threatening activity. Moreover, forced evacuations pressed people to sell their animals on the cheap.[11]

> [We had harvested cotton in our barns.] The soldiers came and burned them down. We moved to another house. They burned down that house, too. . . . I had a large backyard with animals. The fire killed fifty chickens of mine. Three cows and sheep grazing on the prairie survived. But we had to sell them to sustain ourselves.
>
> (Mother Berfo from Siirt, quoted in Yağız et al., 2012: 167–168)

The animals that could not be sold had to migrate to cities along with their humans. In numerous reports from the 1990s and early 2000s, sheep, goats, and cattle were mentioned as roaming in city centers, much like stray dogs or cats (see, for example, Kurban et al., 2008: 241, 228–229). A report published by the Regional Development Administration, a directorate under the Office of the Prime Minister, stated that in the early 2000s 41,000 animals (goats, sheep, cattle) were living among 110,000 humans in Siirt, one of the 14 district centers in the Kurdish region. In Batman, the ratio was 83,000 to 195,000 (Başbakanlık Bölge Kalkınma İdaresi, 2006). Hence, one of the many results of military conflicts is the destruction of ecologies – that is, from a human point of view, particular ties that people have to their environment. Here, I do not want to revive a nostalgic view of the old forms of husbandry. The term 'ecology' does not mean 'natural' or more 'authentic', but connotes a temporary balance between different entities, and it is not necessarily exempt from social hierarchies, violence, or exploitation. Yet, one point has to be underlined nonetheless: Military conflict has deprived people of their animals and destroyed the social fabric in the region. The testimonies from that period often indicate how human suffering was entangled with the suffering of animals during the forced displacements.

> The state banned the use of common pastures. Mountains are infested with landmines. If we cannot take our animals to high plateaus, we can't feed them. Here [in this village], we had 100–150 sheep and goats per household, and all got destroyed. The [Turkish] state made us destitute. Those in the worst condition had to join to the paramilitary guards [*korucu* in Turkish] and fight on the side of the state. . . . In 1993, we bought 400 sheep and goats from Iraq. Our plan was to fatten them up for a couple of months and sell them for slaughtering. [One day], my brother, a nephew of mine, and I were on the prairie to herd the animals. Around 7 or 8 o'clock in the evening, they started firing mortars at us . . . 10–15 animals perished, some got lost. After that night we couldn't use the prairie anymore. My father sold the animals for half their price.
>
> (Sehbaz Enç from Uludere, quoted in Yağız et al., 2012: 299)

Such violence directed against other beings is not exceptional. In other countries, states have also habitually resorted to violence in order to expand or maintain their territorial control. The history of states and colonial expansion is littered with instances of animal murder as a means to overcome resistance. There are numerous cases documented in Afghanistan, Vietnam, Eritrea, Ethiopia, Iraq, and other parts of the world of the killings of 'enemies' animals (see Barker, 1968; Grau, 2004: 135; Sorenson, 2015: 26–27). The western frontier of the United States in the 19th century represents one of the best cases, where a coordinated and premediated attack against the ecosystem could "wipe out the enemy". Particularly between 1860 and 1880, whites started killing bison by the millions. In a speech to the Texas legislature in 1877, General Phillip Sheridan said of the hunters:

> These men have done . . . more to settle the vexed Indian question than the entire regular army has done in the last thirty years. They are destroying the Indians' commissary; and it is a well-known fact that an army losing its base of supplies is placed at a great disadvantage. Send them powder and lead if you will; but for the sake of lasting peace let them kill, skin, and sell until

the buffalo is exterminated. Then your prairies can be covered with speckled cattle and the festive cowboy, who follows the hunter as a second forerunner of an advanced civilization.

(cited in Robbins, 2002: 210)

As a result, "the vast herds of bison on which the Plains Indians had depended for much of their livelihood would die violent deaths and make room for more manageable livestock" (Cronon, 1991: 93, also 213–220; and Smits, 1994). In their place, the landscape was re-organized in favor of cattle. Relations of production, mobilities across space, and biological resources were altered with this violent rupture, as a result of which natives (and bison) became minorities.

Despite differences in time and space, the *modus operandi* of the military apparatus in each example bear a resemblance, in which 'the enemy', as a population, are deprived of their means of subsistence and uprooted from the ecological networks they have developed over time. New relations and new forms of life have replaced what has been destroyed. In other words, the destruction of 'inefficient', 'dangerous' populations made room for other populations. In the section below, I will trace this shift by looking at statistical data from the 1990s to today.

Reshaping regions

Since the 1990s, concerns of security and efficiency have guided large-scale changes in the agricultural sector. One of the parameters that might offer an insight into their scale is the movement in the quantities of different animal species in different parts of Turkey. The two main mechanisms studied in this chapter – that is, military conflict and anti-market policies – have created fluctuations in the number of goats and sheep; 51.1 million sheep and goats had declined to 34 million in 2001, just in 10 years, and further to 29.4 million in 2010 before it increased to 41.5 million in 2014 (tuik.gov.tr, 2015b). I will explain the subsequent sudden increase in the following years below, but first one needs to underline that these are the official figures provided by the State Institute of Statistics of Turkey. Until recently, the animal registration procedure was flawed and incomplete, especially in the regions of conflict. Thus, we can safely assume that the real loss will be much larger than what these figures indicate.

Although we can observe decline as part of a general trend in Turkey, there are significant differences between regions, especially between the West and the conflict-ridden East. When comparing two exemplary districts where intense clashes took place (Batman, Mardin, Şırnak, and Siirt) with those in the West which experienced relative safety during the same period (Tekirdağ, Edirne, and Kırklareli), we can observe a sharp disparity. In the former, the numbers declined from 1.8 million to 860,000 between 1991 and 2001 (52 percent), whereas in the western non-conflict region, the rate of decrease was only 36 percent – from 700,000 to 446,000 (tuik.gov.tr, 2015b).[12]

It would not be entirely correct to assume that the decline in animal quantities is a direct outcome of the conflict. Urbanization, the transition from village production to industrial production, and the reduced tariffs for imports have all their share. They affect different regions and animals in diverse ways. In this regard, the weight of factors is specific to regions and needs to be studied case by case.

To sum up, we can identify two major trends. Between 1990 and 2015, the number of animals first decreased in almost all regions, although at varying degrees. Then, after 2010, they sharply increased again. The important point here is that the rise in numbers does not mean the restoration of the old form of husbandry. Instead, it signifies the shift from one production regime to another. The change in the demographics of cattle will give a better illustration.

In this period, the numbers of cattle (and buffalo) oscillated between 12 and 14 million, with a slight decline and then a rise (tuik.gov.tr, 2015c). One might assume that cattle was not as affected by the new set of agricultural policies, as were goats or sheep. However, in this case, it is misleading to simply look at sums. The extent of the transformation becomes visible only when one teases out the composition by different breeds. Especially after 2000, the so-called inefficient, indigenous breeds drastically diminished in numbers from 6.7 to 2 million. The number of 'high-yield' animals, however, increased almost fivefold, from 1.3 to 6.2 (Table 5.1, below). The state has given incentives, such as subsidies and cheap credit, to import high-yield culture breeds and killed existing schemes supporting indigenous breeds. In other words, even though the quantity of cattle has remained almost the same, the composition of animals has been altered completely.

These changes, like the previous ones, did not influence all geographical areas to the same extent. Regions with big landholders with enough surplus capital and access to cheap credit could invest more in culture breeds and consolidate their position, whereas small producers gradually dropped out of the competition. In the process, regions with a long-standing know-how of husbandry – as well as open spaces and free grasslands, such as in Eastern Turkey – lost against the newly emerging centers which were able to acquire the relatively expensive biotechnological means of industrial production. For example, comparing Eastern Anatolia, traditional free-range cattle breeding sites, with the highly industrialized region of Thrace (Tekirdağ, Edirne, and Kırklareli) in the West, we observe

Table 5.1 Changes in the number of imported culture and indigenous cattle breeding between 1991 and 2014 (tuik.gov.tr. 2015c)

		Imported Culture Breeds	Indigenous Breeds	
	1991	1,253,865	6,685,683	
	1996	1,795,000	5,182,000	
393% increase	2001	1,854,000	4,074,000	70% decrease
	2006	2,771,818	3,405,349	
	2011	4,836,547	2,429,169	
	2014	6,178,757	1,983,415	

that the latter obtained a much bigger share of the imported breeds. Whereas the indigenous breeds dropped by 55 percent (from 440,000 to 199,000) in Eastern Anatolia (Erzurum, Erzincan, Bayburt), the loss was not compensated by the arrival of imported breeds. Their numbers increased by only 85,000 and reached to 121,000. In the same period between 1991 and 2014, imported high-yield breeds in the aforementioned industrial district, reached to almost 400,000 by a three-fold increase (tuik.gov.tr, 2015c). New technologies, along with a new population, enabled enclosed facilities and high volumes of production. Open grasslands in Eastern Anatolia, on the other hand, lost their comparative advantage due to the same set of technologies. In short, the increase in quantities after 2005 did not come from a return to conventional husbandry, but from facilitating a new production regime and the re-organization of space. It involved new actors and was caught in larger financial networks.

One of the key steps for the new relations of production was the 2009 Law on Agro-based Industrial Zones (No. 27402), the concept apparently being inspired by feedlots. The main objective of the zones, as stated in the law, is to enable 'secure' and 'efficient' production in concentrated areas and to combine husbandry with agriculture (Resmi Gazete, 2009). The Southeastern Development Office's report cites as reasons for founding Agro-Industrial Zones the economic inefficiency of conventional livestock production, unregistered animal inflows, and the non-scientific methods of husbandry carried out in urban centers. Not once did the report mention the armed conflict in the region, or why animals were stuck in cities in the first place (Başbakanlık Bölge Kalkınma İdaresi, 2006). "For various reasons", so the report states, "nomadic people arrived" and "hurt the texture" of cities. They polluted the environment and contributed to the deterioration of public health (Ibid.: 5). In order to eliminate these problems, the report recommended to move animals to agro-industrial zones to be opened on the outskirts of cities. Newly built dams would afford the possibility of year-round irrigation and therefore cheap animal feed to sustain these zones. The aim was to establish a 'healthy', 'secure', 'high-yield' production. The plan, however, did not envision villagers to become producers. The zones were not designed for villagers, or their animals. By setting high entry barriers – such as culture breeds, investment in infrastructure, and reliance on external inputs – the objective was rather to create a profitable investment opportunity, with cheap labor already available.

I visited the agro-industrial zone in Diyarbakır in 2013 and conducted interviews with the workers there. One of them, Yahya,[13] told me how the grassland of their village of Övündük, as large as 2,750 acres, had been seized by the state to build this complex. Villagers were not paid any money, because the grassland was a common good and no one held a title deed. After long negotiations villagers were given pre-emptive rights to invest in the zone. For example, 6 acres of land were allotted to Yahya and his 8 siblings. Yet, because they could not come up with the necessary amount of money in due time, their pre-emptive rights were retracted. Among the villagers, no one could take part in the enterprise, so Yahya told me. Such an investment would require start-up money as well as cultural and social capital: personal connections, access to cheap credit, cheap labor, and possibly ties to an import-export company. The first step is to receive a sanctioned blueprint, indicating the technical details of the prospective facility in order to apply for funds. Blueprints are drawn by specialized firms, and even their fee suffice to preclude the majority of villagers from investing. Applying for credit, getting the necessary assurance, following through with legal procedures, importing animals, finding technical assistance, the logistics of distribution, and the like – these all require an entrepreneur's background, rather than that of a villager. Hence, the investors of Diyarbakır's Agro-Industrial Zones consisted of MPs, business persons, or white-collar professionals, but not former villagers. The expense of the buildings alone – those where Yahya was working – amounted to about 820,000 TL in 2013 (circa 400,000 USD then); this sum did not include cattle, wages, or feed.[14]

Yahya sold all his sheep and goats before starting to work there. His monthly salary was 800 TL (395 USD). However, his employer was holding back one year's worth of salary payments. He was practically working for food and minor benefits, similar to the Syrian workers I encountered in other facilities in the same zone. Yahya could not quit, because he had debts and did not want to leave without having received his money. He told me that in the beginning he had no clue how to look after the culture breed animals. The entire setting was different from what he had been accustomed to: the working schedule, the animals' habits, feed, manure, machinery, diseases, and so on. He had to re-learn everything by following the instructions of white-collar experts, especially veterinarians.

The Agro-Industrial Zone may be considered as an investment, adding extra value to the product, but it also serves as a barrier against farmers like Yahya. Moreover, one has to carefully examine the difference between investment and inflating expenses. Under the new regime, the distances that each supply item has to travel are much longer; the carbon emissions increase; grasslands become redundant; and in order to produce animal feed, the soil becomes more and more dependent on agro-chemicals (Emel and Neo, 2011: 79). In 2011, Turkey allowed the import of GMO feed, which might further curtail the farmers' autonomy.[15] Raising livestock independently of climate, soil, existing social relations, and so on in any given setting is accomplished by inserting technical knowledge along with inflated investment requirements. As it is, this tendency seems more like a misguided attempt that will end up in the increase of overall expenses.[16] It also means the increased institutional control of states and corporations over the production process. Animal populations are carefully selected and raised regardless of the particularities of any place, such as climate or the quality of grasslands, and instead become highly dependent on techno-apparatuses. So are the humans. By means of life-enhancing techniques as well as phasing out undesirable beings, entire populations become subjected to certain forms of power. Bio-politics consists of this double-movement.

Conclusion

This chapter has discussed two key dynamics that have changed livestock production in Turkey over the last couple of decades. One of the dynamics has been the shift in the relations of production, while the other has consisted of the military conflict and so-called security concerns in the Kurdish region of Turkey. All along, I have questioned the dominant conceptual framework such as efficiency, investment, free market, and security. In this light, this chapter has illustrated how a new set-up for accumulation through dispossession has taken place in a specific place and time, by considering the normative-regulative policies and the bio-political processes with regard to animals. In short, I have tried to illustrate how economic interventions and violence operated simultaneously to produce the same effect of reducing small-scale livestock production in the Kurdish-majority regions of Turkey.

It might be argued that, with these policies, the state managed to kill two birds with one stone. Certain groups of people have prospered under this regime, while the PKK's logistics have been cut off. From the state's point of view, this might seem like a positive achievement. Yet, the individuals who suffered most from these interventions have congregated in cities such as Diyarbakır, Cizre, Silopi, Nusaybin, Kızıltepe, Siirt, and their surrounding provinces. These are the provinces with the most enduring violence and high levels of poverty and unemployment. In 2015, militants in many Kurdish cities blocked roads and dug ditches to prevent the security forces from reaching the towns, declaring autonomy. The Turkish army crushed the rebellion with disproportional ferocity. My aim here has been to describe the events preceding these clashes. Destroying ecosystems has made rural residents vulnerable, unsafe, and prone to further violence. It is crucial to study the link between capital accumulation, cheap food supplies to cities, and the millions of people affected in order to understand the fault lines within Turkish society.[17]

Another point emphasized here has been the connection between the organization of space and power. I have demonstrated how certain relations between trees, soil, people, animals, and water were obliterated and re-arranged to achieve greater levels of capitalization, by implementing anti-market mechanisms. At times, soft tools such as credit schemes, subsidies, or certificates have promoted the change, but physical force was also frequently utilized. As a result, life forms that are more suitable to mass-industrial production (entrepreneurs + high-yield breeds) were let to thrive, while others became a threat and were terminated altogether.[18]

Notes

1 An earlier version of this chapter was published in Turkish, *Toplum ve Bilim*, 2016.
2 Here, it is necessary to define the concept of security from the very beginning. Evidently, security is a very contested concept. Securing certain groups or territories might lead to the de-securitization of others. For example, the 'security' which was endorsed by the Turkish military during the 1990s has resulted in the devastation of many villages, particularly Kurdish ones. Kurds have lost their means of subsistence, were forced to migrate to bigger cities and therefore have become more insecure (Akın and Danışman, 2011; Yağız et al., 2012; Baysal, 2014). Some even have become expendable in the process. Their graves are still unknown. Thus, every discussion of

security should be sensitive to class, gender, and ethnic segregations, and should treat the term as a governmental device rather than an abstract category (Aradau 2007; for a more detailed discussion see Huysmans and Squire, 2009; Butler, 2010).

3 The claim that developed countries have become self-sufficient is also questionable. Multi-national corporations own large tracts of lands in Africa or Latin America, which they basically use as plantations to feed consumers in wealthier countries. This is not an entirely new phenomenon. The statement about the West being completely self-sufficient in terms of food (Kazgan, 1988: 260) misses the larger networks of oil, finance, and even the extent of transnational food chains that sustain the so-called developed regions of the world. In this regard, the concept of self-sufficiency demands a more nuanced investigation.

4 See Bryant, 1996 on how efficiency was instrumentalized as a colonial endeavor; see also Madra et al., 2002, and Mitchell, 2002 for a discussion of the fabrication of the 'economic principles'.

5 Their term of gestation is about 9.5 months, and procreativity lasts for about 4 years.

6 Feeding livestock exclusively with corn harms the animals. Their rumen is evolved to digest mainly fodder and hay. This is why they are given sponge in industrial production: to trick the rumen into functioning as usual (Shiva, 2006: 78). Also, protein might be derived from various sources; as is well known, the cause of Mad Cow Disease was feeding cows with animal proteins coming from the remnants of industrial meat production.

7 If the opportunity costs of land, water, and other inputs (such as oil) are taken into account in a different manner, calories obtained from animals – as opposed to, say, from a vegetarian diet – do not seem particularly efficient (see Robbins, 2002: 205–206; Singer and Mason, 2006: 58, 122, and 130–132 for alternative calculations). The so-called efficiency is rather facilitated by a coordinated discount on inputs and labor as well as from the tax money (subsidies) channeled into industrial production.

8 For a more detailed discussion on agricultural transformations over a longer time period, see Toprak, 1988; Oral et al., 2013; Özbudun, 2006; Keyder and Yenal, 2013; Sönmez, 2013.

9 The AKP government soon realized that this would cost them the popular vote. They drastically reduced the share of direct income and put back the old and intricate support systems. Nonetheless, compared to European countries, the rate of subsidies to GDP was kept low (see Keyder and Yenal, 2013; Oyan, 2013 for a detailed discussion).

10 I will not dwell on the political conflict and its larger implications (see Bruinessen, 1992; Yeğen, 2006; 2011; Jongerden, 2010; Marcus, 2009; Baysal, 2014).

11 It is important to underline here that the conflict has had (at least) two sides. The more organized violence was exercised by the Turkish state, but there are also testimonies against the PKK confiscating animals by force (for example Yağız et al., 2012: 220–221).

12 'Native' is the classification used by the state. It includes those breeds that have lived in that region for 3 generations and longer.

13 Not his real name. (Interview with Yahya, 2013).

14 Entrepreneurs had to take over half of the financial costs. The EU and the Turkish state were to cover the other half, in which the EU's share was 75 and the state's 25 percent.

15 There is ample evidence indicating that GMO food/feed had already been circulating in Turkey since at least the beginning of 2000. Out of 51 feed samples collected in 2011, right after the law allowing GMO feed, 50 were found to be genetically modified (Evin, 2011, citing the research carried out by Selim Çetiner). This chapter does not intend to delve into the debate of whether GMO is hazardous. Yet, due to these biotechnologies, the relations of production and the spatial arrangements of production shift to favor certain groups over others.

16 Another logic of efficiency based on different premises would have applied other criteria for what type of animal was to be raised in which region. For example, cattle needs long grasses to feed, as its tongue is evolved to wrap and pull out the grass. This is cattle's basic anatomy. Yet, because of the hot climate, Turkey's pastures are typically covered with short grasses (except for a few regions), which are better suited for sheep or goat. Regardless, subsequent governments in Turkey supported the import of culture breeds based on the logic of an entirely different production regime.

17 Moreover, the PKK survived, and the Kurdish issue still remains on the agenda.

18 As we have seen, constructing this new set-up has required a push from powerful actors, but so does sustaining it on the consumption side. The demand for meat has to be manufactured as well. Turkey has an enormous market potential in this regard, because, in comparison to EU countries or the US, Turkey's annual *per capita* meat consumption is very low. When parliament passed the Law on Agro-Industrial Zones in 2009, the annual red meat consumption amounted to only 12 kg per person – extremely low when compared to 62 kg of EU average (Mustafa Albayrak, Chair of the Union of Meat Producers [*Kırmızı Et Sanayicileri ve Üreticileri Birliği, ETBİR*], cited in milliyet.com.tr, 2009). Therefore, increasing the volume of meat consumption is seemingly at the top of the agenda, as is propagated in the language of growth and improvement of health. As a matter of fact, even imperial/national desires are associated with a carnivore diet. In order to promote meat, Fazlı Yalçındağ, the Chair of the Butchers' Federation [*Türkiye Kasaplar Federasyonu*], claimed that the world was ruled by meat eaters, and not by herbivores (cited in Kanat, 2012).

References

Akın, R. C. and Danışman, F. (2011). *Bildiğin Gibi Değil: 90'larda Güneydoğu'da Çocuk Olmak*. Istanbul: Metis.

Aradau, C. (2007). Law Transformed: Guantánamo and The "Other" Exception. *Third World Quarterly* 28(3), pp. 489–501.

Aysu, A. (2016). Interview, 3 January, Mardin.

Barker, A. J. (1968). *The Civilizing Mission: The Italo-Ethiopian war 1935–6*. London: Cassell.

Barut, M. (2001). *Göç-Der Raporu*. Istanbul: Göç-Der.

Başbakanlık Bölge Kalkınma İdaresi. (2006). *Organize Hayvancılık Bölgeleri Kurulması Raporu*. T.C. Başbakanlık Güneydoğu Anadolu Projesi Bölge Kalkınma İdaresi Başkanlığı.

Baysal, N. (2014). *O Gün*. Istanbul: İletişim.

Bryant, R. L. (1996). Romancing Colonial Forestry: The Discourse of "Forestry as Progress" in British Burma. *The Geographical Journal*, 162(2), pp. 169–178.

Butler, J. (2010). *Frames of War: When Is Life Grievable?* London; New York: Verso.

Çağlayan, H., Özar, Ş., and Doğan, A. T. (2011). *Ne Değişti? Kürt Kadınların Zorunlu Göç Deneyimi*. Ankara: Ayizi.

Cronon, W. (1991). *Nature's Metropolis: Chicago and the Great West*. New York; London: W. W. Norton & Company.

Davis, M. and Monk, D. B. (2008). *Evil Paradises: Dreamworlds of Neoliberalism*. New York: The New Press.

De Landa, M. (2000). *A Thousand Years of Nonlinear History*. Cambridge, MA: MIT Press.

Deleuze, G. (2006). *Foucault*. London; New York: Continuum.

Döner, F. N. (2016). Tarımdan Mega Projelere El Değiştiren Topraklar. *Toplum ve Bilim*, 138–139, 67–83.

Emel, J. and Neo, H. (2011). Killing for Profit: Global Livestock Industries and Their Socio-Ecological Implications. In R. Peet, P. Robbins, and M. Watts, eds, *Global Political Ecology*, 1st ed. London; New York: Routledge, pp. 67–83.

esk.gov.tr. (2018). *Et ve Süt Kurumu Tarihçesi*. [online] Available at: www.esk.gov.tr/tr/10097/Et-ve-Sut-Kurumu-Tarihcesi.

Evin, M. (2011). GDO'lu Hayvan Yemi Kullanmayan Yok! *Milliyet*. [online] Available at: www.milliyet.com.tr/gdo-lu-hayvan-yemi-kullanmayan-yok-/gundem/gundemyazardetay/28.12.2011/1481160/default.htm.

Foucault, M. (1979). *Discipline and Punish: The Birth of the Prison*. Harmondsworth: Penguin.

———. (1988). *Madness and Civilization: A History of Insanity in the Age of Reason*, 1st ed. New York: Vintage.

———. (1991). 'Panopticism'. In P. Rabinow, ed., *The Foucault Reader*, 1st ed. London: Penguin Books.

Grau, L. W. (2004). The Soviet-Afghan War: A Superpower Mired in the Mountains. *The Journal of Slavic Military Studies*, 1, 129–151.

Hazine Müsteşarlığı. (2000). *Enflasyonla Mücadele Programı Politika Metinleri Cilt I: Niyet Mektubu, Para Politikası, Ekonomik Kararlara İlişkin Mevzuat*. Ankara: Hazine Müsteşarlığı.

Huysmans, J. and Squire, V. (2009). Migration and Security. In M. D. Cavelty and V. Mauer, eds, *Handbook of Security Studies*, 1st ed. London: Routledge, pp. 161–171.

Jongerden, J. (2010). Village Evacuation and Reconstruction in Kurdistan (1993–2002). *Études Rurales*, 186, 77–100.

Kanat, Ç. (2012). Dünyayı Et Yiyenler Mi Ot Yiyenler Mi Yönetiyor? *Milliyet*. [online] Available at: http://blog.milliyet.com.tr/dunyayi-et-yiyenler-mi-ot-yiyenler-mi-yonetiyor/Blog/?BlogNo=394681.

Karadeniz, F. (2015). İyi Durumdaysak İnsanlar Neden Tarımdan Çekiliyor? *Milliyet Daily*, 17 May. [online] Available at: http://www.milliyet.com.tr/-iyi-durumdaysak-insanlar-neden/pazar/haberdetay/17.05.2015/2060289/default.htm

Kazgan, G. (1988). 2000 Yılında Türk Tarımı: Biyoteknoloji ve GAP Ne Getirebilecek? In Ş. Pamuk and Z. Toprak, eds, *Türkiye'de Tarımsal Yapılar (1923–2000)*, 1st ed. Ankara: Yurt Yayınları, pp. 257–271.

Keyder, Ç. and Yenal, Z. (2013). *Bildiğimiz Tarımın Sonu: Küresel İktidar ve Köylülük*. Istanbul: İletişim.

Kurban, D., Yükseker, D., Çelik, A. B., Ünalan, T., and Aker, T. A. (2008). *"Zorunlu Göç" ile Yüzleşmek: Türkiye'de Yerinden Edilme Sonrası Vatandaşlığın İnşası*. Istanbul: TESEV.

Madra, Y., Özselçuk, C., and Erçel, K. (2002). Bir Tabu Olarak "Ekonomi". *Toplum ve Bilim*, 95, 104–139.

Marcus, A. (2009). *Kan ve İnanç: PKK ve Kürt Hareketi*. Istanbul: İletişim.

Marcus, E. (2005). *Meat Market: Animals, Ethics, and Money*. Boston: Brio Press.

McMichael, P. (2000). *Development and Social Change: A Global Perspective. vol 2: Sociology for a New Century*. Thousand Oaks, CA: Pine Forge Press.

milliyet.com.tr. (2009). Türkiye, AB'nin Beşte Biri Kadar Kırmızı et Tüketiyor. *Milliyet*. [online] Available at: www.milliyet.com.tr/Ekonomi/HaberDetay.aspx?aType=HaberDetayArsiv&ArticleID=1097131&Kategori=ekonomi&b=Turkiye,%20ABnin%20beste%20biri%20kadar%20kirmizi%20et%20tuketiyor.

Mitchell, T. (2002). *Rule of Experts: Egypt, Techno-Politics, Modernity*. Berkeley: University of California Press.

Mutlu, Y. (2015). Diyarbakır'da Köylü, İstanbul'da Kürt: Türkiye'de Yerinden Edilmiş Kürt Gençler. In L. Körükmez and İ. Südaş, eds, *Göçler Ülkesi*, 1st ed. Istanbul: Ayrıntı.

Oral, N., Sarıbal, O., and Şengül, H. (2013). Cumhuriyet Döneminde Uygulanan Tarım Politikaları. In N. Oral, ed., *Türkiye'de Tarımın Ekonomi Politiği*, 1st ed. Ankara: Nota-Bene Yayınları and TMMOB Ziraat Mühendisleri Odası Bursa Şubesi, pp. 71–90.

Oyan, O. (2013). Tarımda IMF-DB Gözetiminde 2000'li Yıllar. In N. Oral, ed., *Türkiye'de Tarımın Ekonomi Politiği*, 1st ed. Ankara: NotaBene Yayınları and TMMOB Ziraat Mühendisleri Odası Bursa Şubesi, pp. 111–130.

Özbudun, S. (2006). *Niçin Dikkulak Oldum? Türkiye Kırsalı Yoksullaşırken. . .* Ankara: Ütopya Yayınevi.

Resmi Gazete. (2009). *Tarıma Dayalı İhtisas Organize Sanayi Bölgeleri Yönetmeliği.* Decree No: 27402. Available at: http://www.resmigazete.gov.tr/eskiler/2017/11/20171125-4.htm

Robbins, R. H. (2002). *Global Problems and the Culture of Capitalism.* 2nd ed. Boston: Allyn & Bacon.

Serhat Kalkınma Ajansı. (2011). *Doğu Anadolu Bölgesi Büyükbaş Hayvancılık Çalıştay Raporu.* Kars: SERKA.

Shiva, V. (2006). *Çalınmış Hasat: Küresel Gıda Soygunu.* Istanbul: Boğaziçi Gösteri Sanatları Topluluğu Yayınları.

Singer, P. and Mason, J. (2006). *The Ethics of What We Eat.* Melbourne: The Text Publishing Company.

Smits, D. D. (1994). The Frontier Army and the Destruction of the Buffalo: 1865–1883. *The Western Historical Quarterly*, 25(3), pp. 312–338.

Sönmez, M. (2013). 90 Yıllık Sermaye Birikim Sürecinin Kilometre Taşları (1923–2013). In N. Oral, ed., *Türkiye'de Tarımın Ekonomi Politiği*, 1st ed. Ankara: NotaBene Yayınları and TMMOB Ziraat Mühendisleri Odası Bursa Şubesi, pp. 11–32.

Sorenson, J. (2015). Animals as Vehicles of War. In A. J. Nocella, II, C. Salter, and J. K. C. Bentley, eds, *Animals and War: Confronting the Military-Animal Industrial Complex*, 1st ed. Lanham, MD: Lexington Books, pp. 17–35.

Steinfeld, H., Gerber, P., Wassernaar, T., Castel, V., Rosales, M., and de Haan, C. (2006). *Livestock's Long Shadow: Environmental Issues and Options.* FAO. Available at: http://www.fao.org/3/a0701e/a0701e.pdf

Toprak, Z. (1988). Türkiye Tarımı Ve Yapısal Gelişmeler. In Ş. Pamuk and Z. Toprak, eds, *Türkiye'de Tarımsal Yapılar (1923–2000)*, 1st ed. Ankara: Yurt Yayınları, pp. 19–35

tuik.gov.tr. (2015a). *ISIC Rev3 Classification, Turkish Imports and Exports.* [online] Available at: www.tuik.gov.tr.

———. (2015b). *Sheep and Goat.* [online] Available at: www.tuik.gov.tr.

———. (2015c). *Cattle.* [online] Available at: www.tuik.gov.tr.

Türkiye'de 50 Baş ve Üzeri Hayvan Varlığı Olan 24 Bin İşletme Var. (2012). *Çiftlik Dergisi.* www.ciftlikdergisi.com.tr/turkiyede-50-bas-ve-uzeri-hayvan-varligi-olan-24-bin-isletme-var.html.

van Bruinessen, M. (1992). *Kürdistan Üzerine Yazılar.* Istanbul: İletişim.

Yağız, Ö., Amca, D. Y., Erdoğan, E. U., and Saydam, N. (2012). *Malan Barkirin: Zorunlu Göç Anlatıları.* Istanbul: Timaş.

Yahya. (2013). Interview. 5 August, Diyarbakır Organize Hayvancılık Bölgesi.

Yeğen, M. (2006). *Müstakbel Türk'ten Sözde Vatandaşa.* Istanbul: İletişim.

———. (2011). *Son Kürt İsyanı.* Istanbul: İletişim.

Yılmaz, Ö. (1999). Nüfus Sayımı Bile Yanlış. *Milliyet Daily.* [online] Available at: http://gazetearsivi.milliyet.com.tr/Nüfus%20Sayımı%20Sonuçları/.

Zeybek, S. O. (2016). Biyo-Politika, Güvenlik ve Anti-Piyasalar: Türkiye'de Endüstriyet Hayvancılığın Seyri. *Toplum ve Bilim*, 138/139, pp. 106–125.

6 Sediment in reservoirs

A history of dams and forestry in Turkey

Ekin Kurtiç

In July 1966, Süleyman Demirel, a well-known Turkish civil engineer-turned-politician famous as "King of Dams,"[1] gave a speech during the groundbreaking ceremony of the Keban Dam. At the time of its construction, Keban was the first and largest hydroelectric dam to be built on the Euphrates in the province of Elazığ, Turkey. Its groundbreaking ceremony was, therefore, an arena for the performance of the developmentalist state that hinged on the promises of improving the life of its citizens through hydraulic infrastructures. In his speech as the then Prime Minister, Demirel pointed to one of the foundational material characteristics of dams:

> Taming a river, incorporating a river into civilized service, into civilization, is no easy task. These mountains will be split in this manner, tunnels will be dug underneath them, and in this great service 5,000 to 6,000 people will work for about four to five years. Then a monument will be erected, and these split mountains and tunnels will all be submerged. When looking at it, one may think that it [the dam] was taken from somewhere else and just brought and placed here.
>
> (BCA 030.01.00.00.16.87.9)[2]

His words are typical of the modernist paradigm of conquering nature. The amount of labor and resources channeled to complete this conquest have become key markers of such "monuments" taming nature. However, Demirel underlined the fact that everything going into the dam's construction would become physically and figuratively invisible due to its own materiality: the dam reservoir would submerge everything under water, including the processes of its own construction. In the end, the dam would appear *as if* it were a *geographic transplant* stripped of all the human and non-human forces that engendered it.[3]

In her study of the oil industry's practices to disengage their work from the specificity and heterogeneity of the production sites, Hannah Appel states that "modularity draws our attention to the productive though ever-incomplete work done in the name of frictionlessness and disentanglement" (2012: 697). What is refracted through Demirel's words is a similar kind of modularity, where the materiality of the infrastructure contributes to the work of disentangling. Dam reservoirs, through the submergence of the landscape, produce a new environment that physically makes invisible the frictions and entanglements embedded in the making of dams. This erasure of the processes and practices that went into its making enables the dammed landscape to appear as natural and seamless. But this is an erasure with limits of its own. The emphasis on the incompleteness of the effects of disentanglement, in Appel's words, gestures to these limits.

The relation between landscape and physical infrastructures is a complex one. As the above quote makes clear, dams – while destroying lands, houses, and roads – simultaneously engender a novel landscape with their reservoirs, roads, and transmission lines. Furthermore, infrastructures are constructed on, and thus embedded in a landscape that is never quite stabilized, as the appearance of dams as seamless transplants suggests. As Andrew Barry argues, "we need to see the earth, its rocks, soil, and water, as integral to the ongoing existence of infrastructure . . . the earth in which infrastructures are embedded is itself neither static nor stable" (2016: 187). The infrastructures therefore inhabit a landscape that is always dynamic and changing, either drastically and/or slowly. Furthermore, the landscape accommodating an infrastructure is not bounded by the exact location where the infrastructure is constructed. The longevity and efficiency of the dam reservoir depend on the amount of sediment that the rivers carry over long distances, which over time accumulates in the reservoir. In short, a dam reservoir is not merely a giant cover concealing the landscape and the processes of its own construction. A dam reservoir does not stand still once completed. Rather, it is an active and dynamic infrastructure that depends not only on the quantity and quality of the stored water, but also on the amount of accumulated sediment. Under the reservoir, therefore, lies not only the landscape marked by life and labor histories, but also the sediment that becomes the infrastructure's own limit (Figure 6.1).

This chapter focuses on the history of dams in Turkey as physical infrastructures that are part of river ecologies carrying not only water, but also sediment. In what follows, I first examine hydraulic infrastructures as a lens to look at the intertwined histories of statecraft and natural resource-making in Turkey. The construction of hydraulic infrastructures has long constituted a process implicated in the state's interests in knowing, measuring, and controlling rivers, thus transforming them into national natural resources. Second, I scrutinize the frictions involved in state- and resource-making beyond the efforts of merely taming nature through technical interventions. The issue of sedimentation in dams, as I will show, brought with it concerns about erosion control and nature conservation. All in all, I argue that dam construction is a site where nature conservation converges with infrastructural expansion.[4] What sedimentation in dams – seems to be hidden under the curtail of the reservoir – in fact reveals is a politics of hydraulic works where infrastructural and environmental interventions collide.

My analysis draws upon 22 months of ethnographic and archival research conducted in Turkey between 2015 and 2018. I have conducted archival research in the Prime Ministry's Republican Archives, the Beyazıt State Library, in the institutional archives of the General Directorate of the State Hydraulic Works,

Figure 6.1 This photo shows the sediment accumulation in the Ayvalı Dam constructed
on the Oltu River, one of the main tributaries of the Çoruh. The soil on both
sides of the water is in fact the sediment accumulated under the dam reservoir. The photo is taken at a location at the upper end of the reservoir where
the seasonal decrease in water level renders visible the sediment that normally
remains "hidden" in the reservoir

Source: Ekin Kurtiç, December 2018

and the obsolete General Directorate of Electric Power Resources Survey and
Development Administration. My ethnographic research took place among local
bureaucrats, engineers, and villagers in Artvin, where dam constructions on the
Çoruh River converge with a watershed rehabilitation project aiming to protect dams against the effects of erosion. Drawing upon this research, this chapter
examines what goes into the construction and maintenance of large dams, with
a specific focus on the problem of sedimentation in reservoirs, which reveals rising concerns about forest and soil conservation as infrastructure becomes to be
understood as a more than technical object. Scholars have studied dams in Turkey as large-scale developmentalist projects that emerge and are sustained despite
recurrent failures in specific political and economic contexts (Çarkoğlu and Eder,
2005; Öktem, 2002); as constructions transplanting novel artificial seascapes and
modern recreational public spaces into arid landscapes (Demirtaş, 2013); as a
site of state-making through the construction of consent for modernization and

development projects (Adaman et al., 2016); as contentious political projects bearing a potential to animate violent conflicts (Jongerden, 2010); as projects that lead to the fall of anti-dam politics due to local populations' spatio-temporal experiences (Evren, 2014); as developmentalist schemes whose global norms of implementation are shifted through the organization of Transnational Advocacy Networks (Kadirbeyoğlu, 2018). This chapter aims to contribute to this literature by focusing on a largely unexamined aspect of dams, that is their very situatedness in a biophysical landscape, and the politics that emerge around the question of their maintenance. By examining how engineers and state officials have been discussing the relation between infrastructure and its environs, this chapter sheds light on multiple forms of state- and natural resource-making that are refracted through hydraulic infrastructure construction.

The making of a "prosperous Turkey" through hydraulic infrastructures

At the heart of the early republican years in Turkey lie multiple, and interwined, forms of construction – that of the nation-state, of natural resources, and of public works. The social and political construction of the nation-state and its nature was deeply entangled with the material construction of railroads, buildings, bridges, dams, and irrigation canals. From this perspective, if the nation in the making was under duress after successive wars,[5] prosperity and revival would eventually arrive by overcoming misery through control over nature and the building of physical infrastructure. A well-known writer, publisher, and the founder of the *Varlık Journal* Yaşar Nabi's short article entitled "İnşa ve Propaganda" (Construction and Propaganda), published in the newspaper *Ulus* (Nation),[6] provides a perfect example of the meaning and value attached to infrastructure. In his column, Nabi (1937) responds to a French journalist's astonishment about the frequency of news published in Turkey about any type of construction works and emphasizes the need for understanding their importance by looking deep into the "nation's psychology." After defining the homeland's most significant problem as neglect and disrepair and stressing the saddening decay of public works, Nabi asks: "Can't you recognize the childlike happiness that this nation, who for too long did not see any positive work [*müsbet eser*] rising in its territory, feels now in face of any small success during the new rising era?" He ends the article explaining the importance of such news with the following words: "Providing examples to the nation by appreciating any constructive [*yapıcı*] and constitutive [*kurucu*] work is the epitome of a positive and creative propaganda" (Nabi, 1937). "Bayındır Türkiye" (Prosperous Turkey) published in *Ulus* on the fourteenth anniversary of the Republic's foundation was one such article. The subheading of the article states that "the trisyllabic word *bayındırlık* encompasses everything, bridges, streets, railroads, postal services, water, buildings, and all that can be given by an advanced civilization" (Bayındır Türkiye, 1937). In Turkish, *bayındırlık* (prosperity) also means public works, as becomes clear in the name of the obsolete *Bayındırlık Bakanlığı* (Ministry of Public Works). The dual meaning of this word shows the close association of public works and infrastructure with a notion of thriving and progress.

In *Building Modern Turkey*, Zeynep Kezer (2015) has shown that the infrastructural and sociospatial reordering was embedded in the making of the nation-state and its citizenry in Turkey. In her words, the construction of public works and the expansion of infrastructural services were to achieve national integration by fulfilling not only a utilitarian function for state operations, but also the symbolic function of creating a sense of national unity in the popular imagination (Kezer, 2015: 160). What went into the making of those infrastructures, however, was not only human labor, capital, interests, aspirations, and desires, but also "natural resources." Key to the process of nation-state building in Turkey, as I show in this section, was the enactment of "natural resources" as knowable, measurable, and controllable entities, together with infrastructure construction and the provision of public works.

Discussions at the 1923 Izmir Economic Congress, which took place seven months before the foundation of the Republic, manifest the close link drawn between the nation-state making and the remaking of the landscape. In his opening speech to the Congress, Mustafa Kemal, the founder and first president of the Republic, signaled the need for human labor and technical tools to transform and develop the national landscape, which would otherwise be nothing more than a mere constraint: "If the homeland was composed of bone-dry mountains, stones, marshy areas, bare lowlands, and of cities and villages, it would be no different than a dungeon . . . The economic reasons and activities are going to transform this country into a developed one, into a paradise." In this expression lies a specific kind of national environmental imaginary, "constellation of ideas that groups of humans develop about a given landscape" (Davis, 2011: 3), which portrayed the unworked national landscape that is untouched by specific human activities as an obstacle preventing human development. In the Republican ruling cadres' environmental imaginaries, "bone-dry mountains" and "bare lowlands," if unworked by human labor and machinery, figured as natural obstacles. Yet, at the same time, according to this imaginary the landscape had the potential to flourish, if enough work was put into it. Mustafa Kemal continued his speech by stressing the need for a combination of labor, tools, and infrastructures to make the national landscape flourish: "We are certainly required to meet the need for manual labor that is lacking for cultivating these wide and fertile lands with technical tools. Besides, we need to make our country a network through railroads and highways on which cars will drive" (Afetinan, 1982: 64–67).

The construction of hydraulic infrastructures became therefore key to the making of the nation-state and its nature. In the early republican years, main concerns were constructing irrigation canals, draining the marshes, preventing floods, and making use of rivers for transportation. It was, therefore, a double struggle *for* and *against* water. In his newspaper column published in 1935, Falih Rıfkı Atay, a republican intellectual and journalist with close ties to the ruling cadres, defined the lack and abundance of water as the two major calamities (*sınat*) of Anatolia. To him, both drought and too much water caused hunger, enabling the characteristics of the landscape to reflect on the "color of the people": "The color of Anatolian peasants in dry places was the yellow of tuberculosis; in watery places it was the yellow of malaria. In the Anatolian struggle for civilization, the biggest work of the republican period will, for sure, be water" (Atay, 1935: 1).

The establishment of state institutions in charge of hydraulic works was a necessary step to expand the knowledge of and control over water. The Ministry of Public Works (*Nafia Nezareti*) was founded in 1920. The central body of the Directorate of the Scientific Hydraulic Committee (*Sular Fen Heyeti Müdürlüğü*) and the regional Hydraulic Works Departments (*Su İşleri Daire Müdürlükleri*) were founded in 1925, under the General Directorate of Public Works (*Umur-u Nafia Müdüriyet-i Umumiyesi*). The Regulation for the Division, Organization and Responsibilities of Water Administration Units, issued on 22 July 1925, stated the aim of these departments as the following:

> Acquiring knowledge about the general condition of waters through a detailed investigation of the country's flowing and still waters and their basins and in this way irrigating the lands and producing electrical power from waters and implementing industrialization wherever possible, improving towns by enlightening the cities and towns through the procurement of economic benefits and helping the development of the industry by benefiting from flowing powers and bringing prosperity to our lands and welfare and happiness to our farmers through the implementation of irrigated agriculture with river waters and revitalizing and giving splendor to our nation, again by draining the deadly and poisonous swamps caused by unruly waters.
>
> (Büyükyıldırım, 2009: 38)

Accordingly, twelve Hydraulic Works Departments were designated. In addition to defining the organization of these administrative units, the regulation also determined the scientific and technical duties of the units, such as preparing maps of the rivers from source to mouth, surveying geological characteristics, illustrating the conditions of forests and the existence of the meadows, marshes, or agricultural lands within the river basins, recording the type and size of sediment in the riverbeds, the existence of irrigable lands, and the type and location of any existent infrastructure working with hydraulic power. The regulation, therefore, paved the way for the advancement and implementation of what is known as "Mobilization of River Reconnaissance Survey" (*Akarsu İstikşaf Seferberliği*) (Büyükyıldırım, 2008). However, due to the lack of a budget, only seven of the envisaged hydraulic departments could be established by 1929 (Özçelik, 2008). Until then, the lack of conducted surveys and budget limitations hindered the development of hydraulic works as planned. Yet, the severe drought of 1928 revealed the need for putting more effort into watering the country.

In response to this severe drought, the effects of which started to be seen in 1926 and which lasted until the early 1930s, the ruling cadres emphasized the need for modernizing agricultural production, alongside the improvement of irrigation infrastructure. In 1929 the *Sular Umum Müdürlüğü* (General Directorate of Waters) was founded; in 1939 it was transformed into the *Su İşleri Reisliği* (Directorate of Water Works). As of 1929, the Parliament initiated the "Great Water Policy," which – together with the railroad policy – in the Journal *Köye Doğru* (Towards the Village) was defined as one of the central pillars of public works for "the growth of the homeland, the nation's welfare and the superiority of Turkish produce in world markets" (Cumhuriyet Devrinde Su Siyaseti ve Köy Davası, 1941: 9). Water survey teams were founded in 1932 for the purpose of preparing reports about the hydraulic regimes of the rivers, which would serve as guides in the construction of hydraulic infrastructures and for river reclamation (Fikret, 1934).

Construction of the infrastructures of the nation-state, therefore, required the physical and symbolic control over its "natural resources." Institutions and regulations were enacted for putting into practice a utilitarian approach to waters through hydraulic infrastructures. As James Scott (1998: 13) has argued, "utilitarian discourse replaces the term 'nature' with the term 'natural resources,' focusing on those aspects of nature that can be appropriated for human use." This utilitarian logic of the high-modernist state, backed by an ideology of belief in science and technology, create an abstract, simplified, and legible nature (Scott, 1998). In his work on the history of environmentalism in Turkey, Dursun (2017: 115) has shown that this utilitarian approach to nature in the Ottoman Empire emerged in the second half of the nineteenth century and converged with conservationist discourses in forest management under the paradigm of a "scientific and rational natural resource" controlled and managed by the state, but according to the principles of economic liberalism. Together with the nation-state building process, Dursun has argued, an authoritarian nationalist and anti-liberal attitude to nature became dominant, while still holding on to the premise of scientific forest management.

My analysis of water-forest entanglements, as reflected in the construction and maintenance of hydraulic infrastructures such as dams, shows how the rise of environmental conservationism in forestry converged with utilitarian policies of water management in Republican Turkey. In other words, a scholarly investigation into the making of dam reservoirs not as a *fait accompli* taming rivers, but as a dynamic infrastructure, allows us to see that forest conservation was pushed for the sake of a better utilization of water resources and maintenance of hydraulic infrastructures. While water became a significant national natural resource that needed to be known, regulated, and utilized thanks to infrastructures such as dams, forests became resources in need of protection for the proper maintenance of these infrastructures.[7] In other words, as Carse (2012: 545) has so eloquently stated, "watershed forests become the infrastructures serving the [Panama] canal." Building on anthropological approaches to resource-making as a socio-political process (Ferry and Limbert, 2008), I suggest that we need to look at how multiple resources are defined and managed in relation to each other. Dams, which have long been analyzed as epitomes of taming and controlling nature, appear to stand in a more complex relationship with natural resource use and nature conservation. In this analysis, while dams have always been technical interventions playing a key role in rendering waters national natural resources to be utilized, perspectives that emphasize water-soil-forest relations have rendered dams into something more than just technical infrastructures. Discussions emergent during the construction of Turkey's first dam in the early Republican era, the Çubuk Dam (1930–1936), exemplify this complex relationship between infrastructure construction, natural resource use, and conservation.

At the time of its construction, the Çubuk Dam was very often in the national newspapers' headlines in a fashion glorifying dams as a path towards building a "prosperous" nation-state. The dam was built 11 kilometers north of Ankara, the new capital, to provide water for drinking and residential, industrial, and agricultural use, as well as to prevent flooding. In addition to these utilitarian functions, the dam acquired symbolic power as a monument to the Republican endeavors of national modernization. In his newspaper column in *Cumhuriyet*, Yunus Nadi defined the dam as a "monument of civilization," which would show "on which paths the new Turkey will have to walk in order to be developed" (Nadi, 1934: 1). The Çubuk Dam became a symbol of conquering Ankara's arid nature through technological interventions, as well as transforming it into a modernized artificial nature with a "transplanted seascape" (Demirtaş, 2013: 17). The dam was depicted as a monument manifesting, what Falih Rıfkı Atay called "the Turkish sword [which] went wherever there was water [and which] is now replaced by a new Turkish technique that brings water to wherever [one lives]." It was a technology to "win over the desert," "revitalize the steppe" and "emancipate" Ankara "from its two longings: water and green" (Atay, 1940: 9). A third, and understudied, aspect of the dams' relations with their biophysical environs that this chapter addresses lies in the nature conservation practices with which hydraulic infrastructures are intertwined.

The production of knowledge about waters and building hydraulic infrastructures were by no means easy endeavors. Officials were concerned with the budget and planning requirements and the limitations of the existing technical expertise. The public, which was supposed to become modernized through the modern spaces built by infrastructures, did not comply with these ideals.[8] Engineers and workers experienced constant challenges while working at construction sites.[9] Similarly, glorified accounts of the Çubuk Dam mask its weak points which in the end rendered it dysfunctional:[10] the dam filled with sediment and became contaminated because of sewage discharge. After 30 years, the Çubuk Dam's dead storage – part of the reservoir's total volume serving to store sedimentation – had been filled up (Çelik, et al., 2012), and after 54 years, 70 percent of the entire reservoir had been filled with sedimentation (Akıncı, 2002, cited in Odabaşı, 2011). The dam became idle as of 1994.

In fact, since the beginning of its construction, intellectuals and forestry experts had raised concerns about the protection of the surrounding forests for a better maintenance of the Çubuk Dam. In their view, proper maintenance of this infrastructure depended on the control of sedimentation in the reservoir and shorten its economic life, if the necessary precautions – such as forest protection and erosion control – were not undertaken. On 11 June 1935, *Ulus Newspaper* ran the headline "The water issue of Ankara is solved," and next to it there was Falih Rıfkı Atay's column entitled "Water." Atay emphasized a less-mentioned aspect of hydraulic works – namely, the relation between water and forests: "The case of water for Anatolia means canals, dams, and forests. Damaging forests in Anatolia means destroying dams, clogging canals. We will rescue our current forests from the hands of profiteers of timber smuggling and will force people to reproduce new forests" (Atay, 1935: 1). In line with the organic view nature and culture hold in the authoritarian nationalist state-making in republican Turkey (Dursun, 2017), Atay (1935, Ibid.) ended his piece by equating water and culture as "the mortar of the structure of Anatolian civilization." Beyond the immediate function that dams serve and the utilitarian logic that prioritizes these functions, Atay's column stressed the long-term maintenance of hydraulic infrastructures that depended on the preservation of forests. The utilitarian logic in managing waters, therefore, converged with a conservationist approach to forests.

Six days after the appearance of Atay's article "Water," Dr. Şeref Nuri, who in 1933 had received his doctorate from a German university with a thesis on the conservation of forests in Turkey, took up the issue where Atay had left off. Stressing the importance of Atay's emphasis on the relationship between water and forest, Nuri stated: "If we want to fix the issue of water in our homeland (*memleket*), the first thing that we will encounter is to arrange our forest policy in an organized and planned way, because forest and water are two interdependent elements" (Nuri, 1935). He then gave a brief scientific explanation about the two major effects of forests on waters: (1) allowing water to infiltrate the soil and gather deeper underground, thus adjusting the water budget, and (2) preventing floods, thus regulating river flows. Noting the lack of scientific experiments about the impact of forests on waters conducted in Turkey so far, he referred to the work of Engler, a Swiss forestry professor, to substantiate his argument on the need to organize forest policy for properly dealing with the issue of water. Quoting from Engler, Nuri wrote that, through their tree roots, forests break, loosen, and mix the soil, thus rendering it more suitable for the infiltration and regulation of water. To him, this was a universal characteristic of the forest soil, and it was also applicable to Turkey. Yet, Nuri also mentioned the need for a site-specific approach to water-forest relations: Nuri claimed that the perspective then dominant in Europe – namely, that forests do not play any important role for big river systems and floods in lower sites – did not apply to the case of Anatolia where floods mostly happened uplands, coming from the higher slopes. In this case, he pointed out the inapplicability of European ideas to Turkey. He summarized: "In our case, floods mostly happen because of the hardening of soil layer due to deforestation and the impossibility of rainfall to infiltrate the soil. Heavy rainfall flows down on a hard layer. This soil not only does not have any benefits or water budget, but also destroys cultivated crops on lower sites and damages the nation's health." Nuri called for the regulation and good management of forests to improve the water policy that, in his own words, is at the center of economic life. In the following decades, as dams proliferated in the country, forestry and agricultural engineers continued to stress their concerns about erosion and sedimentation, emphasizing the harm caused to the national economy and well-being. Such discussions and practices reveal an understudied aspect of dams as monuments of simultaneously conquering, re-creating, and conserving the "natural resources."

Developing natural resources and the nation-state

As of the 1950s, the institutionalization of erosion control within forestry and the proliferation of hydraulic infrastructures in Turkey combined with the rise of a scientific discourse on forest-water relations. This period built on the nationalist state-making project of the early Republican era and added an emphasis on development of both the nation-state and the natural resources.

The institutionalization of erosion control and soil conservation activities in Turkey started in the mid-1940s but became effective only in the late 1950s, "due the lack of education of technical personnel, lack of an institution determined and responsible for this work, and most importantly due to the lack of a sufficient budget dedicated to these activities" (Özkahraman, 1982: 24). In 1956, the Afforestation Group Directorates were founded under the Regional Directorates of Forestry, followed by the foundation of the General Directorate of Forestation and Erosion Control in 1969. In 1957, the Soil Conservation and Pasture Improvement Branch Directorate was founded under the General Directorate of Forestry;[11] meanwhile, the Soil Conservation and Pasture Improvement Practice Group Directorates were established at the local level. These institutions and teams emerged together with the rising scientific expertise on erosion and dry land forestry among the faculty of Istanbul University's Forestry Department in the mid-1950s. Istanbul University graduates became state officials working on erosion control under the forestry institutions. This intertwined process of knowledge production and institutionalization also involved the role of techno-scientific collaborations across national borders, such as a group of engineers who were sent to France in 1953 for a two-year-long training in flood control (Çelik, 1992: 101). Exploring the impact of these cross-border techno-scientific networks is beyond the scope of this chapter. Suffice it to note here that "national" practices of knowledge production and the accompanying interventions in the biophysical environment have been very much embedded in transnational processes.[12]

The 1950s were also a decade when, among other infrastructures such as roads and highways, dams became one of the key markers of the country's economic development and liberalization. After the single-party regime of the early Republican period and its statist policies during the 1930s, by 1946 Turkey had entered a period of multi-party democracy. Pursuing a liberal economic policy, the Democrat Party, which won the 1950 elections, tapped into foreign aid for infrastructural expansion.[13]

In line with the Cold War trend of close relations between the US and Turkey, dam building and watershed management closely followed the US as a model in institutional organization. With the foundation of the General Directorate of the State Hydraulic Works (*Devlet Su İşleri*, henceforth DSİ) in 1954, Turk-ish state authorities closely followed the model of the Bureau of Reclamation in the United States. As Law No. 6200 states, the main objective of the DSİ was to prevent damages caused by surface and underground waters and to ben-efit from them. While the twofold understanding of water as a threatening and also life-giving national "resource" continued, there emerged a new emphasis on "development" in the management of "natural resources," in line with the growing influence of American organizations and institutions in reshaping the scene of hydraulic works in Turkey. In his 1955 book, Gordon R. Clapp, who became Liliental's successor as the Tennessee Valley Authority (TVA) chairman in 1947, defined TVA's approach as a "balanced and comprehensive development of natural resources" (Clapp, 1955: 7). This was enacted in the Tennessee Val-ley "as a good place to apply modern engineering theories to a whole river and combine that development with the inseparable problem of conservation on the land" (Ibid.: 15). Although the TVA's regional development activities started in the 1930s, its expansion into the "developing" and "underdeveloped" world as a model blossomed in the Cold War's geopolitical context (Sneddon, 2015). What was in circulation through this expansion was not only the politics of developing the nation-states and their citizens, but also the idea and practices of develop-ment of natural resources (*doğal kaynak geliştirme*).

Following this rubric of comprehensive natural resource development, the 1950s became the decade when multi-purpose projects and river basin/water-shed management were initiated under the responsibility of the newly founded central state institution, DSİ.[14] In his account of the history of the DSİ, Abdul-lah Demir defined this period as "the era of watersheds (management)" (Demir, 2001: 11). To him, this was an era when "constructing engineering facilities was not enough" (Ibid.). These multi-purpose projects surveyed all the watersheds with their entire resources, such as waters, lands, mines, forests, and human set-tlements (Ibid.). In addition to developing the country through hydraulic infra-structures, hydraulic works, therefore, claimed to provide an integral study and development of all natural resources at the scale of watersheds.

Such endeavors necessitated the inclusion of different engineering practices in hydraulic works. In parallel to the institutionalization of erosion control and soil conservation activities within forestry, and inspired by the US model of the Bureau of Reclamation, the Erosion and Sedimentation Control Branch Office was founded under the DSİ in 1956 to deal with the problem of sedimentation. In his speech addressing the engineers of the DSİ in March 1976, Süleyman Demirel brought up the question as to how hydraulic engineers should reclaim the tribu-taries. Demirel first stressed engineering's role in this endeavor: "We are an insti-tution of engineering. We go there and build a wall around the tributary river. If the river fills the walls [with sediment]. . . . We will rebuild it, elevate the wall" (Demirel, 2005: 250). But then he pointed to the limitations of this engineering solution and suggested that, in the case that building walls around the tributary were not adequate to solve the problem entirely, "you have to turn to look at the degraded balance of nature. There are manifold ways of doing this, ranging from afforestation to building terraces" (Ibid.). He furthermore made clear that this kind of work very well fit their job definition as DSİ engineers: "You cannot say, 'This is not our business.' You are now a 'Scientific Committee for Water Resources' (*Su Kaynakları Heyeti Fenniyesi*)" (Ibid.). His emphasis on the diversity of tasks that DSİ engineers had to perform showcases the recognition of the need for erosion control activities, which brought to the table concerns over nature conservation practices.

Erosion and sedimentation control practices in the upper watersheds did not become effective immediately. While glorifying the multi-purpose dams and "river development program" of the TVA as the "extraordinary advance in man's control over nature," Ward Shepard, who served in the United States Forest Service between 1913 and 1930 and then became the Chief Conservationist of the Indian Office in charge of land policies, wrote in his 1945 book *Food or Famine* that "the great remaining defect of our flood-control planning is that it stops at the river-banks and has as yet developed no significant means of controlling floods where they originate on the land surface of the entire river drainage basin" (Shepard, 1945: 141). Unless such comprehensive and unified river and land control was put into practice, Shepard stated, "the destructive runoff and erosion that are destroying our soils and wrecking our rivers will also annihilate the billions of dollars of investment projected in dazzling but in the long run abortive engineering structures" (Ibid.: 142). Similarly, the need for land use planning and control was still a concern in 1980s Turkey. In his presentation at the Symposium on Land Use Planning in Dams' Upper Watershed, organized in 1987 by the Turkish Association for Nature Conservation (*Türkiye Tabiatını Koruma Derneği*), Necmettin Çepel, an esteemed forestry professor, explained the relationship between hydraulic infrastructures and their surrounding landscape in the following words:

> In a country, developing a planned land use order and its rigorous implementation can be considered as the basic principle for its economic development. . . . For that, it is necessary to know very well the relationship between soil, water and plants, and to evaluate this knowledge very well in the process of making the plans. This subject is of utmost importance for the land use plans in the upper watershed of dams. Because dams, at first sight, appear as mere technical objects storing water and serving the country's agricultural and industrial development. However, for dams to fulfill their functions – in addition to being outstanding construction technology – land use methods in their close and distant environs matter significantly. The reason for that is the fact that the water quantity and quality stored in dams depends on the ecological character of the upper watersheds – that is, the very source where the water comes from.
>
> (Çepel, 1986: 17)

Çepel's quote is a perfect manifestation of the foresters' understanding of dams as more than just technical infrastructures. To better understand this perspective and its implications, I suggest, we need to go beyond the understanding of dams merely as epitomes of a "high-modernist ideology" (Scott, 1998). This critical reconsideration is necessary not only for the recognition of their human and non-human costs, but also to reveal the discussions among the engineering community about dams' own fragility and complexity.[15] By taking dam construction as a practice constituted through multiple forms of expertise, this chapter shows how engineers and state actors tried to deal with "nature," which they cannot fully control by simply building dams. Neither engineers nor state officials easily dismissed the social and material phenomena that did not suit their interests or purposes, as Scott (1998) has suggested. Çepel's emphasis on the necessity of seeing dams not merely as technical, but rather as ecologically entangled infrastructures is a manifestation of modern statecraft as a braided attempt of both simplification and handling of complexities; of controlling and being aware of the limits of control; of knowing and not-knowing (Mathews, 2011). Encountering the limits of the control of water through dams, Çepel called for more and planned control of land use in the uplands.

Controlling the upper watersheds that became a concern for hydraulic works was coupled with the criminalization of upland communities and their livelihood practices. This perspective has rendered upland inhabitants and their livelihood practices a concern to be dealt with. In the book *Orman Davamızın Çeşitli Yönlerine Dair İlmi Görüşler* (Scientific Perspectives on Various Aspects of Our Forest Struggle), published in 1951 by the *Türkiye Ormancılar Derneği* (Foresters' Association of Turkey),[16] professors accused mountain communities and their farming and husbandry practices of destroying the forests. Professor Fikret Saatçioglu (1951: 34), who also served as Minister of Forestry between 1974 and 1975, defined a "struggle between communities which settled in mountainous forest areas and the forests [themselves] taking place in all its severity, to the detriment of our forest and lands," harming the water economy and agriculture of the country. Claiming the impossibility of *in situ* development for these communities, his proposal was for the state to conduct a planned resettlement. In the same volume, Irmak (1951: 47) called for a "rational land use" that would prevent farming on the slopes of the mountainous forest areas from causing the destruction of the "water regime," and thus flooding. To him, the solution lay in opening the uncultivated plains to cultivation. Pointing out the "functions that forests have for the national economy," such as the indirect benefits of "regulating the water and preserving the soil," Acatay (1951) brought to the table the proposition to educate peasants about rotational grazing, to improve forage plant breeding, and to transform husbandry into planned and intensive livestock breeding. In all these accounts, forest villagers, their livelihood practices, and animals were depicted as main culprits responsible for the destruction of the forest and the accompanying degradation of the water regime.[17]

The question of how to intervene in the lives and livelihoods of upland villagers continued over the decades. In the first National Erosion and Sedimentation Symposium, organized by the Turkish Committee of UNESCO's International Hydrology Program in 1978, Muhlis Geyik, an official from the obsolete General

Directorate of Afforestation, indicated forest clearings and unplanned grazing as the main reasons for erosion and lamented Turkey's lack of industrialization necessary to lead people's migration from mountainous and forest areas to industrial centers, thus making these places suitable for afforestation (Geyik, 1978). For Geyik, the issue was therefore less about planned resettlement, and more about aiming at industrialization, with the expectation that it would attract people from mountains to industrialized centers.

Meanwhile, the *Güneydoğu Anadolu Projesi* (GAP; Southeastern Anatolia Project) – begun in the 1970s with the construction of big hydraulic infrastructure projects and expanded in the late 1980s and early 1990s into an integrated regional development project (Özok-Gündoğan, 2005) – became the central hydraulic infrastructure project transforming the river waters of the Kurdish region through a planned total of 22 dams. In the late 1980s, foresters started to push for the inclusion of forestry practices, such as afforestation and erosion control activities, within the GAP, on the premises that the long-term success and maintenance of the entire project in the face of sedimentation and flooding depended on such forestry activities (e.g., Boydak, 1986; Hızal, 1989). It seems that their suggestions started to be put into practice to a certain extent as late as in 1998, through the *Yeşil Öncü Gap Ağaçlandırması* (Green Precursor GAP Afforestation Project) co-organized by Istanbul Technical University and the Turkish Military for the purposes of protecting dam reservoirs from sedimentation. As Özok-Gündoğan (2005) has shown, development policies enacted in the Kurdish region and the state's aim to know, control, and surveil Kurdish people through GAP are organically linked to the trajectory of the war between the Turkish military and the Kurdistan Workers' Party (*Partiya Karkerên Kurdistanê*, PKK). Similarly, in the context of the Kurdish region, along with the emphasis on greening the dams' watersheds, the afforestation project aimed at greening the image of the military, as one of the faculty organizers of the project put it: "Another important point is that this project once again stressed the fact that the Mehmetçik [referring to the soldiers of the Turkish military] who is accused by certain groups for conducting environmental and human massacres while confronting the PKK is attached by love to the stones, land, verdure and university youth of its country" (WebItu, n.d.). Hence, dams and the accompanying environmental interventions unfold in historically and politically situated contexts, predicating on political premises and resulting in power differentials.

Rehabilitating the watersheds

As of the 1990s, erosion and sedimentation control works in dammed watersheds converged with the rise of so-called watershed rehabilitation projects. In line with the global rise of the concept of Integrated River Basin Management (IBRM) in the 1990s – which reemphasizes a "holistic," "comprehensive," and "integrated" approach to water management at the level of the river basin/watershed and calls for a participatory management paradigm (Molle, 2006) – watershed rehabilitation projects started to be implemented in Turkey over the past two decades.

These are the Eastern Anatolia Watershed Rehabilitation Project (1992–2001), the Anatolian Watersheds Rehabilitation Project (2005–2012), the Çoruh River Watershed Rehabilitation Project (2012–2019), and the Murat River Watershed Rehabilitation Project (2012–2018), officially separated into three generations. While all of these focus on the rehabilitation of the so-called degraded natural resources, the first generation only targeted the upper watersheds, while the second generation also included the lower watersheds and the objective of reducing water pollution. The third, current generation of projects has added the aim of proliferating the use of "sustainable energy resources." Concepts such as "sustainability," "participation," and "integrated management" lie at the core of this new type of watershed management projects.

In the Çoruh Watershed, therefore, the convergence of dam construction, erosion control, and natural resource conservation practices currently takes place in a specific historical and political context, where terms such as IRBM, sustainability, and participation received prominence, as they are also being questioned. The Çoruh River, located in the northeastern corner of Turkey, is currently being dammed with ten bumper-to-bumper hydroelectric plants on the main river and five more on its tributaries. Except for the Green Belt Afforestation Projects in Dam Basins, the *Çoruh Nehri Havzası Rehabilitasyon Projesi* (ÇNHRP; Çoruh River Watershed Rehabilitation Project) has been the main forestry and soil conservation project implemented to prevent sedimentation in the dam reservoirs with the collaboration of the General Directorate of Forestry, the State Hydraulic Works, the General Directorate of Agricultural Reform, and the Special Provincial Administration, and is partly funded by a loan received from the Japan International Cooperation Agency (JICA). The project claims to simultaneously rehabilitate the infrastructural, biophysical, and social landscape with a threefold purpose: to prolong the economic life of the dams, to prevent the degradation of natural resources, and to enhance the socio-economic condition of the rural population.

My ethnographic research among foresters, agricultural engineers, and villagers involved in the implementation of the ÇNHRP in Artvin, one of the three provinces of this watershed, shows that this is not simply a linear trajectory towards better integrated, coordinated, and participatory watershed forestry activities. Project designers and implementers stress that *integrated* management means the coordination of activities of various state institutions, which are in charge of different aspects of watershed management, within one comprehensive project. However, coordination *among* institutions does not necessarily mean a better organization *within* each institution. A striking example is the structural change recently made to the organization of the general directorates. In 2011, as the name of the ministry changed from the Ministry of Environment and Forestry to the Ministry of Forestry and Water Affairs, the General Directorate of Afforestation and Erosion Control was closed down and re-organized as Head of Departments under the General Directorate of Forestry. This restructuring has led to a significant decrease in the number of personnel working in afforestation and erosion control, both at the central and local levels. In 2011, within the scope of this restructuring, a new General Directorate for Combating Desertification and Erosion was founded, in a way replacing the former General Directorate of Afforestation. However, as my informants in the Ministry of Forestry have confirmed, this new General Directorate is organized only at the central state level, basically for purposes of policy-making, and does not have any regional and local branches to actually implement afforestation and erosion control works.

Furthermore, the ÇNHRP's participatory approach to natural resource rehabilitation is paradoxically backed by the phenomenon of increasing out-migration, with an accompanying depopulation and decrease in farming and husbandry in the region. The project's master plan defines a "vicious circle between natural resource destruction and rural poverty" (Japan International Cooperation Agency: Pacific Consultants International, 2004: 7), where poverty leads to the excessive use of already scarce natural resources, and thus to their destruction, which in turn aggravates poverty. Yet, the master plan also mentions out-migration from the mountain villages and the resulting decrease in "social pressure" on pasture land (Ibid.: 8) referring to this effect as an important potential for the implementation of rehabilitation work. In response to my question of whether any restrictions on grazing or farming in the mountains was put in place for the purposes of land use control, one of the main project implementers, a forestry engineer, replied while we were driving through the villages to see the accomplished project activities: "Look at these lands; there isn't too much husbandry left here. So there isn't a need to restrict anything." Hence, the villagers' participation in the watershed rehabilitation became, paradoxically, easier due to out-migration and depopulation.

Lastly, these watershed rehabilitation projects are being implemented in an era when socio-environmental movements are on the rise in Turkey. The publication of these rehabilitation projects in shiny magazines and brochures, as well as on websites, demonstrates how they underpin the government's self-proclaimed "greenness." As an attempt to respond to the critiques of socio-environmental movements, which have been opposing to the ecologically destructive policies of the Justice and Development Party government that has significantly relied on big infrastructure and construction projects (Paker, 2017), in many speeches and public addresses, President Erdoğan has recalled the number of trees planted during his party's rule as an indication that the government takes care of nature (Erdoğan: 4 milyar 70 bin adet fidan diktik, 2018).[18] His claims of being the most genuine environmentalist (*çevrecinin daniskası*) have recently manifested through a novel political campaign of building *Millet Bahçeleri* (Nation's Garden) all around the country. Governmental attempts to respond to the critique by trying to co-opt a notion of environmentalism lead to the dismissive attitude of the environmental activists in Artvin, who have been fighting for 25 years against a mining project planned to be installed in the mountains right above the town center, towards these natural resource rehabilitation works. Upon hearing about the implementation of the ÇNHRP while she and I were talking about dams in the region, one of the leaders of the anti-mining movement in Artvin commented: "Which watershed are they going to rehabilitate? Is there any more watershed left?" She referred to the destructive impact of numerous dams transforming the entire river into a succession of dam reservoirs sustained by huge concrete walls. To her, under these conditions, one could no longer talk about a watershed.

Concluding remarks

My analysis of dam building reveals the practices not only of conquering, but also of recreating and conserving nature, which are intimately connected to processes of state-making in Turkey. Dam reservoirs have long been framed by their proponents as monuments to modernization and development, taming and silencing unruly waters, which is only possible at the expense of significant social and ecological destruction (McCully, 1996). Moreover, as Demirel's words in the opening paragraph of this chapter reveal, the reservoirs' physical characteristic of submerging landscapes has made dams appear *as if* they were disentangled from site-specific complexities. A closer look at the discussions and practices among the engineering community, state officials, and intellectuals, however, shows that dams, like any other physical infrastructure, are intimately embedded in the biophysical landscapes of which they are a part.

This chapter, therefore, presents an inquiry into the ways in which the question of maintaining dam reservoirs against erosion and sedimentation has shaped the entangled history of water and forest management in Turkey. This is a sedimented history of nation-state making, developmentalism, and rising authoritarianism, over the course of which water-forest relations have become embedded in new political meanings and governmental practices. I argue that, as the process of hydraulic infrastructure building converted water to a "national natural resource" to be known, controlled, and tapped into, at the same time dams started to be framed as more than technical infrastructures dependent on the forests' protective capacities.

When dam construction proliferated in the country after World War II, US institutions of water management and their framework of watershed management served as a model for hydraulic works in Turkey. The practices of dam building and the protection of watersheds against erosion and sedimentation, in this period, converged within the register of "developing natural resources." At the same time, foresters' emphasis on the importance of conserving upland mountains and forests for the protection of "water resources" was coupled with the criminalization of mountain and forest communities as destroyers of the water regime.

In contemporary context, watershed rehabilitation projects in Turkey are one of the most important sites where the governing of the water-forest nexus materializes. Focusing on the integrated Çoruh River Watershed Rehabilitation Project, this chapter concludes that this historical trajectory does not reveal a linear development towards better integrated, coordinated, and participatory watershed forestry activities. Rather, changes in the institutional organization, long-lasting processes of rural transformation and depopulation, and the rise of socio-environmental movements, as well as the government's self-proclaimed "greenness" all constitute factors that complicate the official discourse about a comprehensive and integrated management of the watersheds and the dams that inhabit them.

Notes

Author's Note

I am grateful to the anonymous reviewers and to the editors Ethemcan Turhan and Onur İnal for their invaluable comments on this chapter. I would like to thank Ajantha Subramanian, Tarık Nejat Dinç, Dilan Yıldırım, and the members of the Political Anthropology and Political Ecology Working Group at Harvard University for their insightful and constructive comments on the earlier version of this chapter. This work was supported by a İstanbul Friends of ARIT Research Fellowship, ZEIT-Stiftung Ebelin und Gerd Bucerius Trajectories of Change Fieldwork Grant, Weatherhead Center for International Affairs Dissertation Completion Fellowship, and Cora Du Bois Summer Fellowship. I am deeply grateful to archivists, librarians, villagers, bureaucrats and engineers, and scientists who generously shared their insights and helped me in realizing this research. All translations from Turkish to English are mine.
1 Demirel served as the General Director of State Hydraulic Works (1955–1960), then became Turkey's Prime Minister (intermittently between 1965–1993), and later President (1993–2000).
2 As is customary in the historiography of Turkey, I will use the acronym BCA for the archival sources from the Başbakanlık Cumhuriyet Arşivi (Republican Archives of the Prime Ministry), although the name of this archive has changed to the Republican Archives of the Presidency in July 2018.
3 In her study on the creation of built environments in Turkey through infrastructures of water, Demirtaş (2013) calls built reservoirs "transplanted seascapes," which introduce bodies of water into the arid landscapes in order to transform nature, align it with purposes of modernization and democratization. Here, using Demirel's speech as an illustrative example, I emphasize another characteristic of reservoirs as transplants – that is, their effect of creating a landscape *as if* it was frictionless and seamless.
4 Scholarly work on nature conservation practices in Turkey is now on the rise. For an examination of the co-production of scientific forestry and the state through the discourses and practices of rendering verdure an index of civilization, see Özkan (2013). For a study of the making of wetlands in Turkey and wetland conservation as a relation of care, see Scaramelli (2018). See also Soylu, Akbulut and Adaman (2017).
5 The wars in question are the Balkan Wars (1912 and 1913), World War I (1914–1918), and The War of Independence (1919–1923).
6 The newspaper *Ulus* (Nation) belonged to the Republican People's Party, the single ruling party in power in early republican years.
7 Forest conservation for water regulation and hydraulic infrastructure protection was not always the only practice. Görcelioğlu (2004) has explained that in the United States of the 1940s one of the strategies to protect dams through watershed management was to minimize or totally destroy vegetation, on the premise that this would prevent the use of water by plants, thus allowing more water to be collected in dams. Yet, this was limited to a few pilot implementations in Arizona, and its expansion to other areas was stopped due to environmentalists' reactions.
8 Belge's short piece published in *Ulus* newspaper is a clear example of how people's usage of the recreational area upset the project of creating "modern" citizens through "modern" built environments. In his column, Belge complained: "The state made everywhere spotlessly clean. But the people of Ankara, . . . will almost wash their clothes

and their children's stool there. Because they cook their food there. Just in front of the dam. They bring everything with them, their barbecue, their coal, their child's swing . . . Sometimes a piece of newspaper swims in the water. The state as well as the people should fight against this sort of people who somehow cannot leave the tribal mentality behind. The dam was not built for this kind of people" (Belge, 1937: 2).

9   Sadettin Acar, a civil engineer who began to work in the *Su İşleri Reisliği* (Directorate of Hydraulic Works) in 1947, indicated some of the difficulties in the maintenance of hydroelectricity infrastructures. At that time, Acar was responsible for examining projects of potable water provision and hydroelectricity projects in small towns. Since there existed no one with the relevant expertise, Acar was put in charge. He remembers that these small-scale hydroelectric plants "got filled and clogged in the first flood. Only a few of them served their purpose for a while; later on, they all disappeared" (Demir, 2006: 90).

10 Blackbourn (2006: 230) has referred to siltation behind the dam walls as an indication of self-destruction: "All dams defeat their own purpose in the end."

11 At that time, the Directorate was under the Ministry of Agriculture. The Ministry of Forestry was founded much later, in 1969.

12 As Carse (2014) has noted, dissemination of ideas about water-forest relations in the United States started in the mid-nineteenth century and the early twentieth century witnessed the proliferation of scientific work on the influence of forests on hydrology. This influence was named as "forest hydrology" in 1948 and the idea of watershed forests gained prominence (Carse, 2014: 42–43). Forsyth and Walker (2012) have also shown that, although ideas about the forests' role in hydrology are still debated, they widely circulate among scientific and expert communities and related state institutions.

13 For a thorough study of meanings and practices of highway construction in Turkey after World War II, see Adalet (2018).

14 Together with the foundation of DSI, 26 watersheds were designated in Turkey and water management started to be organized at the river basin level. On the making of river basins/watersheds as a socio-political process and on the social life of river basins, see Swyngedouw (1999) and Molle (2009).

15 For another study that recognizes the fragility of dams from a different angle – that is, the disasters caused by them – see Huber, Gorostiza, Kotsila, Beltrán, and Armiero (2016).

16 In 1951, the society became the Foresters' Association of Turkey. Source: www.orman cilardernegi.org/index.asp

17 For a detailed account of the criminalization of forest communities in republican Turkey, see Özkan 2013, pp. 79–90.

18 For arguments opposing such statements, see: www.dw.com/tr/t%C3%BCrkiyede-fidanlar-art%C4%B1yor-ormanlar-azal%C4%B1yor/a-47080666?maca=tr-Twitter-sharing; www.evrensel.net/haber/310760/akp-orman-varligini-biz-arttirdik-diyemez

Bibliography

Afetinan, A. (1982). *İzmir İktisat Kongresi, 17 Şubat-4 Mart 1923*. Ankara: Türk Tarih Kurumu.

Acatay, G. (1951). *Orman Davamızın Çeşitli Yönlerine Dair İlmi Görüşler*. Ankara: Türkiye Ormancılar Cemiyeti.

Adalet, B. (2018) *Hotels and Highways: The Construction of Modernization Theory in Cold War Turkey*. Stanford, CA: Stanford University Press.

Appel, H. (2012). Offshore Work: Oil, Modularity and the How of Capitalism in Equatorial Guinea. *American Ethnolojist*, 39(4), pp. 692–709.

Adaman, F., Akbulut, B., and Arsel, M. (2016). Türkiye'de Kalkınmacılığı Yeniden Okumak: Hes'ler ve Dönüşen Devlet-Toplum-Doğa İlişkileri. In C. Aksu, S. Erensü, and E. Evren, eds, *Sudan Sebepler: Türkiye'de Neoliberal Su-Enerji Politikaları ve Direnişleri*, 1st ed. İstanbul: İletişim Yayınları, pp. 291–312.

Atay, F. R. (1935). Su. *Ulus*, 11 June, p. 1.

Atay, F. R. (1940). Bozkır. *Köye Doğru*, 8, p. 9.

Barry, A. (2016). Infrastructure and the Earth. In P. Harvey, C. B. Jensen, and A. Morita, eds, *Infrastructures and Social Complexity: A Companion*, 1st ed. London: Routledge, pp. 187–198.

Bayındır Türkiye. (1937). *Ulus*, 29 October, p. 5.

Belge, B. (1937). Ankara'nın Boğaziçisi. *Ulus*, 26 July, p. 2.

Blackbourn, D. (2006). *The Conquest of Nature: Water, Landscape, and the Making of Modern Germany*. New York: W. W. Norton & Company.

Boydak, M. (1986). Güneydoğu Anadolu Projesi (GAP) ve GAP'ta ormancılığın yeri. *İstanbul Üniversitesi Orman Fakültesi Dergisi*, 36(2), pp. 75–93.

Büyükyıldırım, G. (2008). Akarsu İstikşaf Seferberliği ve Genç Cumhuriyet'in Mühendisleri. *Türkiye Mühendislik Haberleri*, 44, pp. 42–46.

———. (2009). Türkiye'de Su Mühendisliğinin Öncüsü, Türkiye Cumhuriyeti'nin 2. Bayındırlık Bakanı: Süleyman Sırrı. *Türkiye Mühendislik Haberleri*, 453, pp. 36–39.

Çarkoğlu, A. and Eder, M. (2005). Developmentalism *alla Turca*: The Southeastern Anatolia Project (GAP). In F. Adaman and M. Arsel, eds, *Environmentalism in Turkey: Between Democracy and Development?*, 1st ed. Ashgate: Aldershot, pp. 167–184.

Carse, A. (2012). Nature as Infrastructure: Making and Managing the Panama Canal Watershed. *Social Studies of Science*, 42(2), pp. 539–563.

———. (2014). *Beyond the Big Ditch: Politics, Ecology, and Infrastructure at the Panama Canal*. Cambridge, MA: MIT Press.

Çelik, H. E., et al. (2012). İstanbul'un İçme Suyu Barajlarının Sedimantasyon Problemi ve Çözüm Önerileri: Alibeyköy Barajı Örneği. *İstanbul Üniversitesi Orman Fakültesi Dergisi*, 62(2), pp. 113–127.

Çepel, N. (1986). Barajların Yukarı Yağış Havzaları İçin Arazi Kullanım Planlarının Ekolojik Esasları. *İstanbul Üniversitesi Orman Fakültesi Dergisi*, 36(2), pp. 17–27.

Clapp, G. R. (1955). *TVA: An Approach to the Development of a Region*. Chicago: University of Chicago Press.

Cumhuriyet Devrinde Su Siyaseti ve Köy Davası. (1941). *Köye Doğru*, 21, pp. 7–9.

Çelik, H. E. (1992). Havza Islahının Türkiye'deki Tarihsel Gelişimi. *İstanbul Üniversitesi Orman Fakültesi Dergisi*, 42(1–2), pp. 95–112.

Davis, D. (2011). Introduction: Imperialism, Orientalism, and the Environment in the Middle East: History, Policy, Power, and Practice. In D. Davis and E. Burke III, eds, *Environmental Imaginaries of the Middle East and North Africa*, 1st ed. Athens, OH: Ohio University Press, pp. 1–22.

Demir, A. (2001). *DSİ ve Su Tarihi*. Ankara: DSİ Vakfı Yayınları.

Demirel, S. (2005). *Süleyman Demirel: Bir Ömür Suyun Peşinde*. Istanbul: ABC Medya Ajansı.

Demirtaş, A. (2013). Rowing Boats in the Reservoir: Infrastructure as Transplanted Seascape. In P. Pyla, ed., *Landscapes of Development: The Impact of Modernization Discourses on the Physical Environment of the Eastern Mediterranean*, 1st ed. Cambridge, MA: Harvard University Press, pp. 16–36.

Dursun, S. (2017). The History of Environmental Movements and the Development of Environmental Thought in Turkey, 1950–1980. In H. Petrić and I. Žebec Šilj, eds,

Environmentalism in Central and Southeastern Europe: Historical Perspectives, 1st ed. London: Lexington Books, pp. 111–131.

Erdoğan: 4 milyar 70 bin adet fidan diktik (2018). *T24*. [online] Available at: https://t24.com.tr/haber/erdogan-4-milyar-70-milyon-adet-fidan-diktik,741340 [accessed on December 15, 2018].

Evren, E. (2014). The Rise and Decline of an Anti-Dam Campaign: Yusufeli Dam Project and the Temporal Politics of Development. *Water History*, 6(4), pp. 405–419.

Ferry, E. E. and Mandana, E. L. (2008). *Timely Assets: The Politics of Resources and Their Temporalities*. Santa Fe: School for Advanced Research Press.

Fikret, T. (1934) Cumhuriyet Rejiminde Su İşleri ve Su Kuvvetinden Elektirik İstihsali. *Nafıa İşleri Mecmuası*, 2, pp. 14–16.

Forsyth, T., and Walker, A. (2012). *Forest Guardians, Forest Destroyers: The Politics of Environmental Knowledge in Northern Thailand*. Seattle: University of Washington Press.

Geyik, M. (1978). Dağlık havzalar erozyon kontrolü çalışmalarında karşılaşılan sosyoekonomik sorunlar. In *I. Uusal Erozyon ve Sedimentasyon Sempozyumu Tebliğleri, Ankara, 25-27 April, 1978*, Ankara: DSİ Yayınları, pp. 208–219.

Görcelioğlu, E. (2004). Havza Amenajmanının Dünü, Bugünü, Yarını. *İstanbul Üniversitesi Orman Fakültesi Dergisi*, 54(2), pp. 1–13.

Hızal, A. (1989). Güneydoğu Anadolu Projesi (GAP) ve ormancılık. *İstanbul Üniversitesi Orman Fakültesi Dergisi*, 39(1), pp. 77–88.

Huber, A., Gorostiza, S., Kotsila, P., Beltrán, M. J., and Armiero, M. (2016). Beyond "socially constructed" Disasters: Re-politicizing the Debate on Large Dams Through a Political Ecology of Risk. *Capitalism Nature Socialism*, 28(3), pp. 48–68.

Irmak, A. (1951). Orman ve Ziraat Topraklarından Rasyonel Bir Şekilde Faydalanmalıyız. In Orman Fakültesi Profesör ve Doçentleri, eds, *Orman Dâvamızın Çeşitli Yönlerine Dair İlmî Görüşler*, 1st ed. Ankara: Türkiye Ormancılar Cemiyeti, pp. 45–48.

Japan International Cooperation Agency: Pacific Consultants International. (2004). Türkiye Cumhuriyeti Çoruh Katılımcı Havza Rehabilitasyonu Master Plan Çalışması. Available at http://open_jicareport.jica.go.jp/pdf/11744075_01.pdf

Jongerden, J. (2010). Dams and Politics in Turkey: Utilizing Water, Developing Conflict. *Middle East Policy*, 17(1), pp. 137–143.

Kadirbeyoğlu, Z. (2018). Waterproof Development? Impact of Advocacy Networks on Anti-Dam Movements in India and Turkey. In S. T. Jassal and H. Turan, eds, *New Perspectives on India and Turkey: Connections and Debates*, 1st ed. London; New York: Routledge, pp. 182–194.

Kezer, Z. (2015). *Building Modern Turkey: State, Space, and Ideology in the Early Republic*. Pittsburgh: University of Pittsburgh Press.

Mathews, A. S. (2011). *Instituting Nature: Authority, Expertise, and Power in Mexican Forests*. Cambridge, MA: MIT Press.

McCully, P. (1996). *Silenced Rivers: The Ecology and Politics of Large Dams*. London: Zed Books.

Molle, F. (2006). *Planning and Managing Water Resources at the River-Basin Level: Emergence and Evolution of a Concept*. IWMI Comprehensive Assessment Research Report 16. Colombo, Sri Lanka: International Water Management Institute.

Molle, F. (2009). River-Basin Planning and Management: The Social Life of a Concept. *Geoforum*, 40(3), pp. 484–494.

Nabi, Y. (1937). İnşa ve propaganda. *Ulus*, 20 July, p. 5.

Nadi, Y. (1934). Tekemmül Yolunda Bir Şehrimiz: Ankara. *Cumhuriyet*, 5 March, p. 1.

Nuri, Ş. (1935). Orman ve Su. *Ulus*, 17 June, p. 6.

Odabaşı, B. (2011). *Rezervuarlarda Sediment Birikimin Önlenmesi ve Rezervuar Ekonomik Ömrünün Uzatılması*. MA. Istanbul: Istanbul Technical University.

Öktem, K. (2002). When Dams Are Built on Shaky Grounds: Policy Choice and Social Performance of Hydro-Project Based Development in Turkey. *Erkunde*, 56(3), pp. 310–325.

Özçelik. C. (2008). Türkiye'de Su Hizmetleri ve Su Hukukunun Gelişimi. *DSİ Teknik Bülteni*, 103, pp. 10–22.

Özkahraman, İ. (1982). *Atatürk'ün Doğumunun 100. Yılında Ağaçlandırma, Erozyon Kontrolü ve Mera Islah Çalışmaları*. Ankara.

Özkan, H. (2013). Cultivating the Nation in Nature: Forestry and Nation-Building in Turkey. Ph.D. New Haven, CT: Yale University Press.

Özok-Gündoğan, N. (2005) 'Social Development' as a Governmental Strategy in the Southeastern Anatolia Project. *New Perspectives on Turkey*, 32, pp. 93–111.

Paker, H. (2017). The 'Politics of Serving' and Neoliberal Developmentalism: The Megaprojects of the AKP as Tools of Hegemony Building. In F. Adaman, B. Akbulut, and M. Arsel, eds, *Neoliberal Turkey and Its Discontents: Economic Policy and the Environment Under Erdoğan*, 1st ed. London; New York: I. B. Tauris, pp. 103–119.

Saatçioğlu, F. (1951). Memleketimizde Ziraatle Orman Arasındaki Mücadele ve İç İskan Zarureti. In Türkiye Ormancılar Cemiyeti, ed., *Orman davamızın çeşitli yönlerine dair ilmi görüşler*. Ankara: Türkiye Ormancılar Cemiyeti, pp. 34–45.

Scaramelli, C. (2018). "The Wetland Is Disappearing": Conservation and Care on Turkey's Kizilirmak Delta. *International Journal of Middle East Studies*, 50(3), pp. 405–425.

Scott, J. (1998). *Seeing Like a State: How Certain Schemes to Improve the Human Condition Have Failed*. New Haven, CT: Yale University Press.

Shepard, W. (1945). *Food or Famine: The Challenge of Erosion*. New York: The Macmillan Company.

Sneddon, C. (2015). *Concrete Revolution: Large Dams, Cold War Geopolitics, and the US Bureau of Reclamation*. Chicago: University of Chicago Press.

Soylu, C., Akbulut, B., and Adaman, F. (2018). The Political Economy of a Conservation Plan: The Case of Ulubat Lake. In M.F. Göçek, ed., *Contested Spaces in Contemporary Turkey: Environmental, Urban and Secular Politics*. London; New York: I.B. Tauris, pp. 257–280.

Swyngedouw, E. (1999). Modernity and Hybridity: Regeneracionismo, and the Production of the Spanish Waterscape, 1980–1930. *Annals of the Association of the American Geographers*, 89(3), pp. 443–465.

WebItu (n.d.). [online] Available at: https://web.itu.edu.tr/~arslanc/personal/gap/bilgi-2.htm [accessed on January 2019].

7 Informalization of waste regimes

The entanglement of urbanization, poverty and waste in Ankara

Gül Tuçaltan

Informality, or unrecognized forms of value creation, is a well-visited and acknowledged axis of research outside the Global North when waste is concerned (see Millington and Lawhon, 2018: 4–7). Encompassing a diverse array of perspectives and empirical analyses from many geographical contexts, a considerable number of studies have examined the characteristics of labor and economic activities that are not controlled or regulated by the state. These studies have, for the most part, not addressed the complex set of relations that shape the emergence and transformation of informal practices. Instead, they consider informality as a pre-given, static characteristic and a sectoral component of waste systems in the Global South.

How, then, has informality become an inherent characteristic of waste systems? To what extent do changing political, economic, institutional and material (waste composition, technologies, land where waste is dumped) processes shape the informalization of waste systems? This is the question I address in this research by treating the informality of waste systems as a "social and historical process, rather than as a sector" (Castells and Portes, 1989; Meagher, 1995: 264; see also Gidwani, 2015). Drawing on the notion of waste regimes (Gille, 2012: 29), I argue that informalization cannot be considered independently of the "social institutions and conventions that not only determine what wastes are considered valuable, but also regulate their production and distribution". Informality, as a dynamic phenomenon, is at the intersection of entangled political, economic, institutional and material (waste composition, technologies) processes – and their unintended consequences.

For instance, informality of waste regimes could be shaped by political-economic changes that redefine the industrial production and societal consumption patterns (e.g., a shift from production of import-substitute to export-oriented consumption goods). It could be informed by related material transformations (e.g., changes in the volumes of production, introduction of new packaging technologies, increasing volumes of waste and recyclable materials in waste composition, such as plastic and paper products). Inflows of unskilled labor (e.g., rural-to-urban migration under decreasing levels of agricultural employment) and institutional responses (e.g., introduction of new responsibilities for the authorized actors) could shape officially recognized and unrecognized technologies of waste management (e.g., municipal open dumping, informal recycling). To show how changes in these entangled factors of transformation jointly provide the context for waste to become a source of livelihood for the urban poor, I focus on the processes involved in the informalization of waste systems in Ankara from the 1930s to the 2000s. I show how socio-spatial, material and economic links between the urban poor, waste and urban growth emerged in the course of rapid urbanization and population growth, which, then, re-shaped the city's waste regimes.

The remainder of this chapter is structured in four main sections. In the next section, I present the major conceptual and analytical considerations of informality in waste systems. I introduce the concept of waste regimes, which allows informality to be situated in the multi-faceted relations that underlie the production of value from waste. The second section points out the key moments and historical processes of political, economic and institutional change, as well as their unintended consequences, and the material ramifications that resulted in the informalization of waste systems in Ankara. In the third section, I discuss the key findings' potential contribution to the literature and the explanatory power of my theoretical approach. I conclude with policy recommendations and suggestions for further research.

The question of informality and waste regimes

The current research on the nature of informality of waste systems has three major axes. The first locates the question of informality within the broader processes of capitalist accumulation. Recognizing informal practices as an income-generation strategy, these studies are critical to the consideration of the terms, formal and informal, as a binary opposition (see Varley, 2013; Gidwani, 2015) and point to the intertwined nature of formal and informal practices. They demonstrate how the state intentionally causes the vulnerability of informal waste pickers for the sake of sustained material flows and capital accumulation (O'Hare, 2017). They show how local informal practices enable capital and material flows (Corwin, 2018). They describe the production and dispossession of informal recyclers within capitalist urbanization processes (Gidwani, 2015; Gidwani and Reddy, 2011; Inverardi-Ferri, 2018) and the potentialities for class struggle they cause (Campbell, 2018).

The second major axis of research concerns the living conditions and everyday life of informal waste workers, which they discuss in relation to the social relations of power. These studies focus on everyday politics and the conditions of informality within the scope of class, racial or cultural politics (Fredericks, 2014; Njeru, 2006; Parizeau, 2015; Wittmer and Parizeau, 2018). They show how power relations at various scales, for example, between local and national government authorities, local workers and communities, reproduce the everyday politics of informality (Bjerkli, 2013, 2015; Cornea et al., 2017; Gutberlet, 2016). They show how power dynamics, not only between formal and informal actors, but also within informal economies, underlie the stigmatization, exclusion and marginalization of informal laborers (Adama, 2014; Gutberlet and Jayme, 2010). It is widely recognized that international development aid and international actors (Myers, 2005; Adama, 2014), privatization (van Horen, 2004; Samson, 2010) and municipal attempts to achieve technological change (Bjerkli, 2013; 2015; Demaria and Schindler, 2016; Schindler and Kishore, 2015) have reshaped the positionality and attitude of formal local actors towards waste, urban space and informality. In particular, privatization in the Global South has fragmented waste service delivery and redefined the role of informal actors (van Horen, 2004), and the nature of this redefinition is gendered, racial and class-based (Samson, 2010).

A third group of scholarship discusses informality and the question of integrating it into formal systems. Bottom-up strategies such as self-organization and cooperative building (Gutberlet, 2008, 2012; Lawhon et al., 2018; Medina, 2007, 2011; Nunn and Gutberlet, 2013), social entrepreneurship (Nzeadibe, 2013; see also Oteng-Ababio, 2018 for a consideration of gender roles) appear as recognized strategies for bridging informal and formal activities. Discussions of the public policy relevance of integration into formal systems often refer to the weakness of the extant institutional and technological systems and the need for appropriate system designs and techno-managerial solutions (see Alemu, 2017; Simatele et al., 2017; Wilson et al., 2006).

Whether discussing informality in relation to capitalist accumulation, everyday life or the effectiveness of public policies, all three clusters of studies relate the current occupational patterns of informal workers to individual segments of reality, namely the economy, social relations of power and institutions. Doing so limits their ability to provide sufficiently interrelated accounts of power mechanisms, along with the historical, political, economic and socio-spatial contexts and material and technological processes through which informality is produced and reproduced. I argue that such treatments of informality in waste systems do not grasp the complex relations between waste, poverty and urbanization – in contexts where the heterogeneity (see Gidwani, 2015; Jaglin, 2014; Lawhon et al., 2018; Tuçaltan, 2017) of socio-technical configurations through the co-existence of a variety of officially recognized/formal and unrecognized/informal actors and technologies is a dominant feature of waste systems. I further argue that this leads to the development of unsuitable and/or insufficient policy and planning responses.

To address this gap, I use the concept of waste regimes as developed by environmental sociologist Zsuzsa Gille (2007, 2010). Like "resource regimes" (Young, 1982), which act as social institutions that produce or delimit social value for specific natural resources, waste regimes define the value attributed to waste. The process of valorization or devalorization happens through a dynamic interplay between (Gille, 2012: 29):

- changes in political economy and public policy and related waste production patterns,
- materiality (i.e., composition of waste) and dominant technologies of waste management,

- representations and discursive constructions of waste by various actors (i.e., efficiency/inefficiency, usefulness/uselessness, order/disorder, gain/loss, clean/dirty), and
- power relations between formal and informal actors.

The outcome of this interplay between these factors of transformation is the formal or informal, intentional or unintentional distribution of advantages and disadvantages in society through the production of rules, rights and decisions, which are always open to redefinition (Gille, 2010). In short, the waste regimes approach allows us to analyze informality as a social and historical process as it emerges and becomes inherent to waste systems.

In order to present the processes of informalization in relation to the entangled factors of transformation presented above, the remainder of this chapter presents a historical and processual account of the waste systems in Ankara, Turkey, between the 1930s and the mid-1990s. Ankara is an effective entry point for discussion as a very dynamic, rapidly changing, growing and urbanizing city. It is an exemplary case concerning the co-existence of formal and informal forms of waste treatment in the context of a strong state and legal-regulatory tradition. An activist who was actively engaged in organizing informal waste workers commented: "To understand waste in Ankara is to understand Turkey" (personal communication on September 18, 2014). In this respect, Ankara offers important clues about the informalization of the waste sector in Turkey's metropolitan areas. In the next section, I present the methodological and empirical considerations that drove the selection and scope of the case study alongside the time period covered in this research.

Methodological and empirical considerations

Since the 1930s, various regulatory measures have been introduced in Turkey to directly or indirectly manage municipal solid waste flows, but it took until the 2000s before concrete interventions that enabled transformation in formal waste management systems were put on its urban agenda. After gaining momentum with the start of the European Union (EU) negotiations in 2005, the Turkish government initiated a mass legislation process and adopted the EU legislation in the field of waste management. According to Ministry of Environment and Urbanization (MoEnU, 2016: 145), between 2003 and 2016, the proportion of the population served by waste disposal and recovery facilities to the municipal population doubled with the nation-wide construction of new facilities and increased from 32 to 65%. In addition to the sanitary landfill construction projects, there are 21 facilities that can generate electricity from methane gas (including electricity generation from depot gas and biomass plants) (Ibid.). A large number of privately-owned waste collection and sorting and recycling facilities has been established. Despite these elements of transformation, it is not possible to refer to a thorough shift in municipal waste systems in Turkey. Informal open dumping practices, unsanitary municipal dumping sites and illegal reclamation of recyclables from landfills still exist to a large degree. Due to the lack of at-source separation and a developed medical waste management system, discarded materials sent to landfills include food, recyclable materials such as plastic, paper, glass, metals, composite wastes, and even medical waste, electric electronic items and furniture. Therefore, informal waste picking, storage, sorting and transfer constitute large parts of Turkey's waste systems.

A closer look at the case of Ankara from the perspective of Turkish waste governance and management and its transformation reveals important parallels between the two. Since its declaration as the capital city of the newly established Republic of Turkey in 1923, Ankara's population has increased more than a hundredfold, even before the 2000s, and the urban area has grown from 280 hectares (Çamur and Yenigül, 2009) to approximately 2.5 million hectares (Ankara Valiliği Çevre ve Şehircilik İl Müdürlüğü, 2016: 49). Prior to the 2000s, sanitary disposal, as well as extended networks for formal recycling of waste, did not exist in Ankara. In the 2000s, Ankara witnessed significant transformations in its waste disposal and recycling sector, namely the development of controlled disposal and energy production systems, the rehabilitation and transformation of its municipal waste dumping site and an increase in the number of actors engaged in authorized recycling and disposal. Nevertheless, there is still no systematic at-source separation of valuable materials in household waste, and recycling laborers, who developed self-governing networks to extract the increasing economic potential of waste much earlier than the formal actors, constitute an important group in Ankara's waste system.

This research focuses on the central districts in Ankara where not only the major socio-technical transformations have taken place, but also the population density and the volumes of waste produced are higher. It presents and discusses the transformation in the eastern corridor, namely the district of Mamak, which was never considered an axis for urban development and has historically been the locus of urban poor and the city's officially designated open waste-dumping sites (see Figure 7.1). The deliberate focus on the period before the transformation of waste regimes between the 1930s and the 2000s – when formal actors rapidly became the dominant presence – is the result of my desire to present a detailed historical analysis of the multi-faceted factors that underlie the processes of informalization.

This case study research was done in three field visits covering a period of eight months between 2014 and 2016 and a short preliminary visit to the field in the late 2013. In the absence of systematic studies of the volumes and composition of waste and statistical data, semi-structured in-depth interviews, focus group discussions and observation constituted the primary sources of data along with the review of secondary sources such as newspaper archives, planning reports and environmental assessments. In the remainder of this chapter, I analyze the waste regimes in Ankara between the 1930s and the mid-1990s, before the rapid entry and increased influence of formal actors. I show how informality became a definitive element in the changing relations between waste, poverty and urbanization in Ankara. Considering when, how and in what ways informalization became a

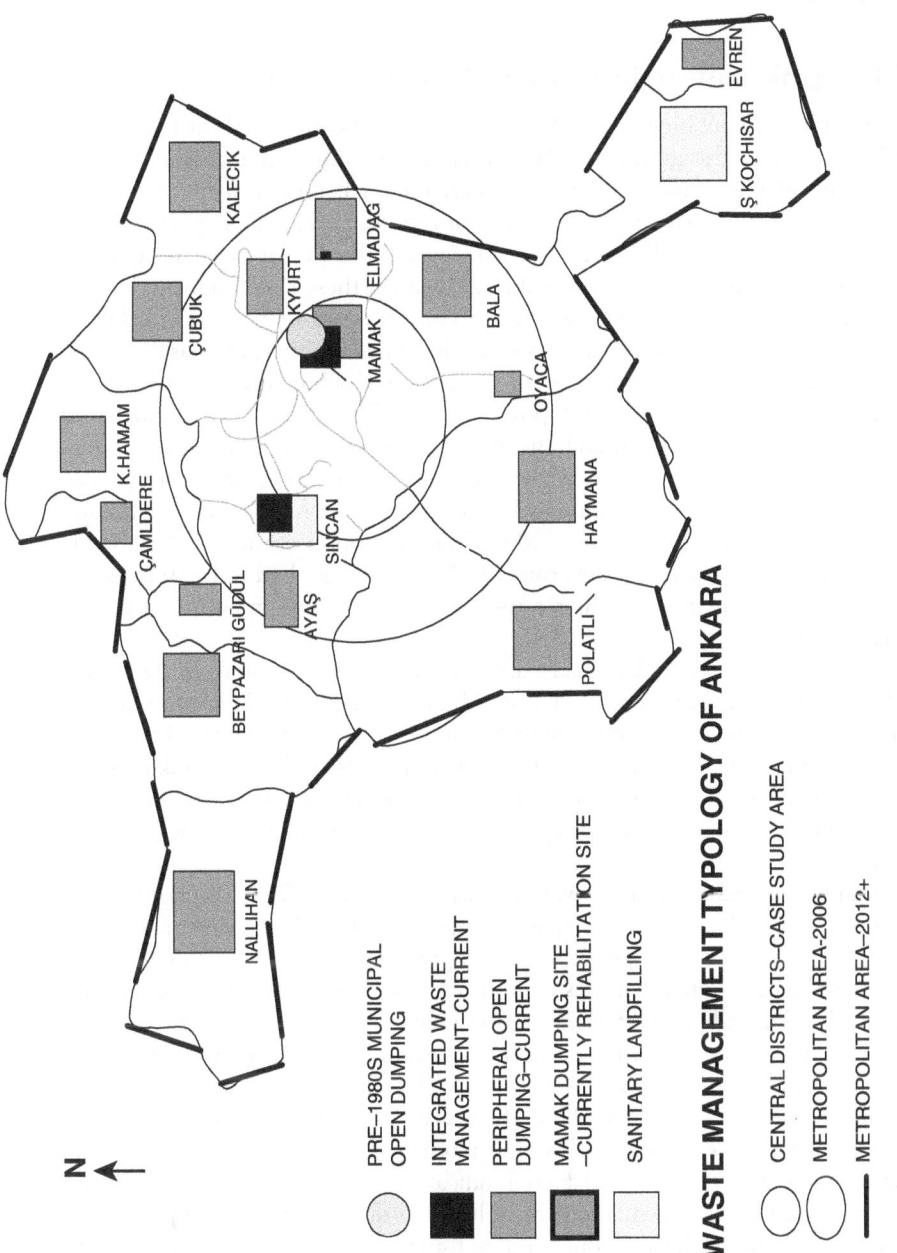

WASTE MANAGEMENT TYPOLOGY OF ANKARA

N

EVREN
Ş KOÇHISAR

KALECIK
ELMADAĞ
KYURT
BALA
ÇUBUK
MAMAK
OYACA
K.HAMAM
ÇAMLDERE
SINCAN
HAYMANA
BEYPAZARI GÜDÜL
AYAŞ
POLATLI
NALLIHAN

PRE–1980S MUNICIPAL
OPEN DUMPING

INTEGRATED WASTE
MANAGEMENT–CURRENT

PERIPHERAL OPEN
DUMPING–CURRENT

MAMAK DUMPING SITE
–CURRENTLY REHABILITATION SITE

SANITARY LANDFILLING

CENTRAL DISTRICTS–CASE STUDY AREA

METROPOLITAN AREA–2006

METROPOLITAN AREA–2012+

Figure 7.1 Ankara's Districts and Waste Management Typology

Source: Figure by the author, based on Ankara Valiliği Çevre ve Şehircilik İl Müdürlüğü (2016)

significant characteristic of Ankara's waste regimes, it is possible to refer to three distinct periods: 1) the organic regime from the 1930s to the 1950s, 2) the regime of volume from the 1950s to the 1980s, and 3) the regime of value from the 1980s to the 2000s.

The organic waste regime from the 1930s to the 1950s

The Turkish Republic was established in 1923 at the crossroads of Europe and Western Asia in the aftermath of World War I. The constitutive cadres adopted a statist development model, which considered municipalities a significant device of the state for implementing its policies at the local level (Tankut, 1990: 114). The Municipal Law (*Belediye Kanunu*) and the Public Health Law (*Umumi Hıfzıssıhha Kanunu*) were both issued in 1930 under these circumstances. Despite the fact that neither of these laws directly concerned waste, both included items related to public health and sanitation. Both defined municipalities as the acting authority for the cleanliness, order and sanitation of public spaces along with the collection, transportation and disposal of waste. Thus, waste was considered to be a public health concern and an issue of urban order and cleanliness (Güler et al., 2001). However, the state's developmentalist approach was mainly based on establishing a national economy by means of a nation-wide industrialization and transport strategy (Keskinok, 2006). With the onset of the world economic crisis in 1929, the national government's limited budget had to be strategically utilized and prioritized. A major consequence was the minimization of investment in urban services (Şengül, 2009: 166).

As the capital city, Ankara had a distinct position as the spatial representation of the state's political project. That led the municipality to direct state investments towards development of public buildings, open spaces, modern boulevards and housing projects symbolizing the new and modern Turkey (Tankut, 1990; Keskinok, 2006). Thus, developing a waste management system was not a priority area for the municipality, except for keeping the streets of the new city clean and orderly.

Limited financial resources were not the only reason for waste not being a major concern. Its material composition also played a role. First, Ankara was a small city with low waste production levels. Second, in the absence of advanced industrial production, urban waste mostly consisted of biodegradable matter and coal ash (especially in winters), so it was a common practice for citizens to dispose of their own waste or bury it in their yards. In addition, the population was still engaged in agricultural activities, and feeding animals with food waste was a common practice. The remainder of the waste was disposed through wild dumping by citizens or the municipality. The municipal archives I was able to access and the interviews I conducted did not indicate any officially designated dumping sites for this period. The municipality fulfilled its duties by means of open dumping far from the settlement zones and by throwing waste down into the valleys in the agricultural lands of southeastern Ankara. Ankara's first dumping site was the village of Mühye, located in İmrahor Valley, and Ankara's waste was dumped in this area until the 1950s. Therefore, in this early period, waste management was an institutional responsibility, but for a municipality that lacked the financial means to manage the task properly.

The regime of volume from the 1950s to the 1980s

Waste's entry onto municipal agendas between the 1950s and the 1980s was due to the changing dynamics of urbanization and urban development. The electoral victory of the conservative, populist and foreign aid and investment-friendly Democrat Party (*Demokrat Parti*; DP) was a definitive moment in Turkish history. Statist economic policies gave way to an import-substitution model based on the domestic production of goods to limit the volumes of imports (Keyder, 1987). This shift in policy and politics had unintended consequences for Ankara's waste management in a number of ways.

The first was the unintended consequence of Turkey's openness to foreign aid mechanisms. Turkey was one of the nations aided by the Marshall Plan, an economic support mechanism provided by the USA after World War II. The stipulations of the aid program dictated that the funds were directed to agricultural modernization and highway construction. Turkish agriculture at the time was heavily based on unskilled labor power. Thus, the introduction of new technologies – the tractor, in particular – and the mechanization of agriculture had drastic effects on its labor dynamics. Masses of people lost their source of income, became surplus labor, and migrated to urban centers with the hope for better occupational opportunities. Defining this process as the urbanization of labor, Şengül (2009: 122) claims that policies concerning rural areas had more drastic implications for urban centers than they did for rural areas themselves in terms of the production of income inequality, informality and clientelism. Neither the national government nor local governments were willing or able to provide newcomers with jobs, housing or infrastructure services. The implications of the mass migration wave included: 1) increasing volumes of waste and 2) the emergence of early recycling and reuse practices by the urban poor.

The population increase and the growth in waste volumes resulted in municipalities' reconsideration of open dumping practices. Nonetheless, this reconsideration was not free from the general budget constraints following World War II. "Far from the city, with no potential economic value and no expectations of future development, owned by the State Treasury, which would mean no further action regarding the expropriation of private property to claim land for dumping, the area was more than appropriate," is how an interviewee summarized the reasons for designating Ankara's eastern corridor, Mamak, as its open dumping destination (personal communication on August 27, 2014). He further explained that the area's hilly topography made its valleys convenient for dumping. Once thrown into the valley, waste was no longer a concern. Guided by this logic, Kartaltepe was selected as the first officially designated waste dumping site. According to the records of Mamak Municipality, the site started operating in the 1950s and remained active until the municipality began to use the Ege site in 1964. Interestingly, in less than two decades, the city decided to close the Ege dumping site in 1978 and designated another. The new official designation for Ankara's waste, the Mamak dumping site, was located in southern Mamak next to the İmrahor Valley, which had been proposed as a forestation area (Dilek, 2006) by Ankara's then-current master plan. With ecological significance for Ankara as a ventilation corridor and a water basin, İmrahor was and still is an ecological asset for Ankara. The municipal site selection criteria that determined the choice of its predecessors were also valid for this site: distant from the urban core, not located in the development corridors of the city and adjacent to a valley.

The second indirect ramification of aid from the Marshall Plan, in the absence of the necessary supply that could meet the housing needs of the migrants, was the rapid *gecekondu* construction and urban sprawl in the 1950s (Erman, 2004). *Gecekondu* literally means "built over night," and refers to squatter housing built quickly and haphazardly by newcomers to Turkey's cities. It became a phenomenon in Turkey due to lack of a public response to the housing need caused by the rapid rural-to-urban migration (see Davis, 2006: 38–57). The factors that directed municipal waste dumping towards Mamak were also what made the area attractive to newcomers who wanted to build *gecekondu*: being located at the peripheral areas of the city on unvalued and undeveloped State Treasury land.

When first selected, both the Kartaltepe and Ege dumping sites were located on undervalued land next to valleys, which are parts of Ankara's water basin. When waste was thrown down the hills into the valleys, it was both physically and psychologically out of sight. Nevertheless, this short-term convenience for the municipality did not last long. In a focus group (personal communication, September 26, 2014), long-term residents of the area suggested that first Kartaltepe in the 1950s, and later Ege in the 1970s became *gecekondu* neighborhoods. The discussion also revealed that municipal dumping at the Ege site continued until the 1990s. This means that waste was being dumped on an already developed *gecekondu* neighborhood.

This close relation between waste and *gecekondu*, which shaped Ankara's urban development at the time, was not reflected in the city planning of the period. Like the local government, the national government had no consistent policy regarding *gecekondu* areas, and it was not until the 1980s that further planning measures regarding *gecekondu* were put on the authorities' agenda. Not considered a development corridor, eastern Ankara was being devalorized by the hands of public actors.

Another significant ramification of political and economic change in the 1940s was the change in waste composition and the emergence of early forms of formal and informal recycling. I have mentioned that Turkey had adopted an import-substitution industrialization model as its core economic policy. In the absence of the necessary industrial infrastructure, the implementation of the import-substitution model in Turkey was based on privatization as well as foreign credit, investment and aid (Boratav, 2015). This weakened the nation's protectionist economic policies based on balancing import and exports (Ibid.: 96). Although a military coup ended the Democrat Party's rule in 1960, the development perspective remained an important component of the Turkish economy until the 1980s (Ibid.: 119–120). This distinct articulation with the world economy resulted in changing production patterns, especially towards the end of the 1970s (Ibid.: 97). The production of goods such as televisions, radios, automobiles, domestic appliances, textiles and office supplies mainly symbolized the consumption patterns of advanced capitalist countries (Ibid.: 121). As a matter of course, the changing production patterns had direct implications for the production and the materiality of waste, which were studied by the Ministry of the Interior in 1969. This nationwide study (see Arun, 1972: 55–56) found that Turkish waste was mainly composed of food waste, paper, bottles, jars, glass, cans, tin plates, bottle tops, animal bones from butcher shops and restaurants, leather, old clothing, rags, sawdust and plastics.[1] The reuse and recycling potential of the items in the waste stream was already being realized within either larger or smaller production processes, or everyday practices, but to a very limited extent due to the absence of systematic collection and separation mechanisms (Ibid.: 64–71).

Arun's (1972) comprehensive analysis of the pre-1970 waste systems in Turkey suggests that salvageable materials in the waste stream were collected and sorted by economically disadvantaged citizens, for reuse or sale. My own empirical research shows that these disadvantaged citizens were mostly newcomers to the city. During my fieldwork in the Tuzluçayır neighborhood, I met an old woman who migrated to Ankara from the Central Anatolian city of Çorum some 50 years ago. She told me that it was a common practice for *gecekondu* residents to salvage useful materials from the dumping site, which they could use or sell (personal communication on September 9, 2014). They did not only dig through the open dumping sites, but also in uncontrolled dumps and trash cans. She specifically referred to plastic and paper as "useful materials." However, she saw coal as the most valuable material. An elected opposition party member of the Mamak Municipal Council, who has been actively working with Halkevleri,[2] confirmed her story. He accompanied me during one of my field visits to the neighborhood, and as a resident there since childhood, he claimed that "everybody used to go through trash for useful stuff" (personal communication, September 9, 2014). In addition to my personal communication, an ethnographic study of the formation of the Ege neighborhood in Ankara refers to similar information regarding the earlier recycling practices (Şentürk, 2015). As the previous section noted, the second officially designated open dumping site of the city is named after the Ege neighborhood. In her research, Şentürk (Ibid.: 81–82) tells the story of Fidan, the daughter of a seasonal agricultural worker who migrated to Tuzluçayır in the 1950s. Lacking sufficient income for food, Fidan – who was already used to rummaging through trash for food for her siblings while her father wandered for work even before moving to Ankara – collected materials from the dump site to build their *gecekondu*.

The regime of value from the 1980s to the 2000s

Between the 1980s and the mid-2000s, informality became a dominant feature in Ankara's waste regime. The findings revealed how three major reconfigurations redefined the nature of these relations revolving around the materiality and technologies of waste and land, and the urban poor. First of all, political and economic change underlying the shift from import substitution to export-oriented industrialization redefined not only the city's industrial production and societal consumption patterns, but also its waste composition.

At the beginning of the 1980s, Turkey was in a serious economic crisis. Towards the end of the 1970s, the import-substitution industrialization model based on foreign aid, credit and investment, in place for almost four decades, was no longer functioning. The major reason for this model's failure was the global economic depression caused by drastic increase in oil prices starting in the mid-1970s. This limited not only production, but also the foreign loan and credit flow to Turkey (Balaban, 2008). The crisis of production had severe implications: increasing inflation rates, devaluation of currency, diminishing real wages and loss of jobs, which, as a matter of course, negatively affected consumers' purchasing power (Boratav, 2015). The economic unrest and its consequences for the labor force manifested themselves in the emergence of a wide range of militant labor movements and the strengthening of labor unions. These caused management and control problems for capital holders and reduced levels of production (Ibid.: 145–146).

The national government's response to the economic and civic unrest was an economic program that was introduced in January 1980. Known as the notorious January 24 decisions, this economic program entirely restructured the Turkish economy through the introduction of the International Money Fund (IMF) and the World Bank (WB) structural adjustment perspectives of the 1970s (Ibid.: 151–152). Instead of attempting to restore failing domestic production and the import-substitution model, the program was based on a complete, even extreme, reorganization of the Turkish economy under the principles of an export-oriented model. Export-led growth implied opening domestic markets to foreign competition in exchange for market access in other countries (Balaban, 2008: 77–81). These so-called reforms, privatization, trade liberalization and the financial deregulation of markets (Aksoy, 1993) significantly transformed the commodity and labor markets (Yeldan, 2006). According to Balaban (2008: 77–78), such a radical restructuring of the economy was not possible under "normal and democratic circumstances."

In less than a year, the existing government was overthrown by a military coup on September 12, 1980. Under the military junta, political parties were banned from politics, and all trade union activities were stopped (Boratav, 2015). A revised constitution was ratified in 1982. Remaining in power for the following three years, the military withdrew from the government in 1983. In that year, general elections were held in a rather undemocratic environment due to the ban on political parties under the military junta. In these circumstances, Turgut Özal and his new, right-leaning Motherland Party came to power in 1983. His administration was a turning point for Turkish economy, politics and policies in every possible segment. Generally associated with his contemporaries Thatcher and Reagan, Özal was a strong advocate of neoliberalism. In consonance with this ideological stand, Özal played a key role in the introduction of neoliberal ideas to the Turkish economy. In particular, his policies on privatization and deregulation were reflected in waste regimes.

In this early stage of neoliberalization, commodity trade liberalization appeared as the major form of integration into global markets (Yeldan, 2006). Both imports and exports increased radically (Yeldan, 2006; Boratav, 2015), and industrial production patterns (Balaban, 2008) and societal consumptions patterns changed. For instance, Turkey started manufacturing PET and plastic products starting in the 1980s for packaging in lieu of metal (İKMİB, 2013). The use of aluminum cans instead of glass in the beverage industry, and the use of corrugated cardboard in packaging, gained momentum in the mid-1980s (İKMİB, 2013). The 1980s also constituted a turning point in consumption patterns. The spread of supermarket chains towards the end of the decade and the establishment of the first shopping malls were two major developments that directly influenced lifestyles and increased the consumption of packaged products (Erberk et al., 2011).

In the absence of statistical data from the period, the evidence from interviews with planning experts who were long-term residents of Ankara (personal communication, September 24, 2014; personal communication, December 12, 2016) found that this transformation in the production patterns spurred the changing material composition of waste and increased the volumes of recycling materials such as rubber, textiles, glass, metals, different kinds of plastic, cardboard, paper, composite waste and bulky waste such as furniture, and electric and electronic waste. The first formal study of the composition of municipal solid waste in Turkey, conducted by TurkStat[3] in 1993 (TÜBİTAK,[4] 2004), validates this information.[5]

The Institutional changes reformulated the grounds for decision-making and implementation mechanisms (Davoudi, 2009). A two-tier municipal system was introduced for metropolitan areas in 1984. Ankara became a metropolitan municipality with a number of district municipalities. The Environmental Law No. 2872 (1983) dictated that waste responsibilities were to be redistributed, with the metropolitan municipality assigned the responsibilities of waste treatment and disposal, and district municipalities assigned the responsibility of waste collection. The municipalities were given the authority to privatize their waste-related responsibilities as public-private partnerships or by subcontracting. My findings indicate the impact of this new ability to privatize. The attempts by two consecutive mayors to enable socio-technical change could only be put on the agenda after this institutional reconfiguration. Although ineffective, the attempts at technology transfer in collaboration with foreign consultants in the 1980s (the construction of a facility that was not compatible with material characteristics of Ankara's waste) and waste planning in the 1990s in collaboration with the private sector indicate the change in the understanding of waste by the municipal actors: from an undesired externality to a resource (Demaria and Schindler, 2016).

While municipal dumping continued officially in Mamak and unofficially in Ege, in 1986, a newspaper report referred to 2,000–2,200 tons of municipal solid waste production per day and efforts by the Metropolitan Municipality to develop responses to its waste problem (Özbilgen, 1986). According to this newspaper report, as early as the mid-1980s, the municipality prepared a plan for the disposal and treatment of the 2,000–2,200 tons of waste produced daily, which was to establish a waste-sorting facility, an incineration facility for electricity production and a transfer station system to improve waste collection. The facility was eventually established. However, it did not operate properly and was left to decay (tbmm.gov.tr, 1985). According to the representative of the Recycling Laborers Association (*Geri Dönüşüm İşçileri Derneği*, personal communication on January 24, 2015) and a former employee of the Municipal Construction Affairs Department of the Ankara Municipality (personal communication on September 24, 2014), its operational failure was due to the improper technology for the high humidity levels of Ankara's waste at the time. Later, the residents of the area started to use the empty buildings as animal shelters.

In the late 1980s, the Ege dumping site was still in operation, and the Mamak dumping site had already expanded towards agricultural lands and become notorious for its odor problem and leachate filling the valley basin. It was not only the valley and its *gecekondu* settlements that suffered the deleterious effects of the dumping site. The malodor from the dumping sites wafted into adjacent sections of the district of Çankaya where the former Presidential Palace was located. Under these circumstances, local elections took place in 1989. Murat Karayalçın from the Social Democratic Populist Party (SHP; *Sosyaldemokrat Halkçı Parti*) won and remained in power until 1994.

In 1991, the first regulation directly concerning waste, namely the By-law on Solid Waste (*Katı Atıkların Kontrolü Yönetmeliği*)[6], was enacted. Subsequently, the Ankara Metropolitan Municipality determined to act on Ankara's waste, started a new initiative to designate a new sanitary landfill area with a proper methane discharge system and a medical waste incineration plant. Acknowledging the already established links between waste, urban development patterns and the urban poor, the metropolitan municipality subsequently initiated a rehabilitation process for the waste dumping sites and *gecekondu* areas in Mamak. The Mamak municipality also prepared reclamation plans for the *gecekondu* areas. In addition, the metropolitan municipality placed a number of bins for paper and glass recycling (Ankara Büyükşehir Belediyesi, 1993), but they did not function due to lack of a systematized infrastructure for recycling and the public's lack of interest. Around this time, the Ankara Metropolitan Municipality, in collaboration with Middle East Technical University, a number of NGOs and the private sector (*Ankara Sistem Planlama*) also prepared plans for a new sanitary landfill and waste treatment facility in Ankara's western corridor. The new site was planned as part of a holistic waste collection system supported by transfer stations located in every district municipality (Sistem Planlama, 1992). The transfer stations were planned to function as storage and sorting facilities for the reclamation of the recycling materials in waste. The remainder of the waste was planned to be sent for disposal (Ibid.). After a comprehensive planning study, the Sincan Çadırtepe area located 55 km west of the city center was selected as the prospective disposal site (Ibid.). Due to budgetary constraints, funding negotiations with the World Bank and the processes of necessary expropriations, the project slowed down and eventually ground to a halt by the time of the local elections in 1994. İ. Melih Gökçek's new administration took no action concerning waste until 2002 when waste management services were subcontracted for 49 years to a private company, Invest Trading and Consulting (ITC).

In the absence of municipal mechanisms to realize the economic potential in the waste streams' recycling materials, the forced migration of ethnic Kurds due to civic unrest in Turkey's eastern and southeastern cities in the 1990s caused labor inflows that constituted a major force in the consolidation of informality in Ankara's waste treatment. An unintended consequence of the conflict between the Turkish army and Kurdish paramilitary groups indelibly redefined the character of Ankara's waste regime at the time. Forced migration and the transplantation of a considerable number of displaced ethnic Kurds to "conflict-free" areas by the Turkish armed forces coincided with the increased amounts of valuable materials in waste. The representative of the Recycling Laborers Association said:

> For these people, the chances of living in the developed areas of the city were zero. They were unskilled for urban jobs. Their languages, color, culture, their everything was different. So, they turned to waste-picking, which does not require skills, and waste, a zero-cost income provider.
>
> (personal communication on January, 24, 2015)

The destination of some of the Kurdish migrants to Ankara, who originated mainly from southeastern provinces of Hakkari, but also from Van and Şırnak, was Türközü, a neighborhood close to the Mamak dumping site, which basically turned into a giant warehouse at the time (Ibid.). The new migrant groups soon got into business relationships with the former migrant groups (Central Anatolians) who had already built warehouses on the Mamak dumping site. Some recycling laborers built their own warehouses in their places of residence. Therefore, unauthorized recycling became a systematic practice for many residents, most of whom lived in *gecekondu* areas near the dumping site. They not only reclaimed waste from the dumping sites, but also collected waste from the bins located in the city center. These early systematic practices continued with no major interference by the public authorities until the early 2000s, when the metropolitan municipality subcontracted its waste management functions to ITC.

Another group, mainly consisting of Kurdish migrants from the city of Gaziantep, also migrated to Ankara in the mid-1990s for economic reasons. The major cause of migration for these recycling laborers, who had been peasants and livestock breeders in their hometowns, was the loss of these jobs. The policy schemes introduced in the field of agriculture in the mid-1990s were based on privatization and the elimination of state economic enterprises (Yörür, 2010: 17). They also included the elimination of existing regulatory mechanisms for agricultural land protection and pricing (Ibid.). Neoliberal rural and agricultural policies were not compatible with the extant agricultural production patterns based on small-scale production, but with large-scale production (Ibid.). This resulted in the loss of jobs and subsequent migration for many small producers, peasants and livestock breeders (Özuğurlu, 2013). These groups arrived in the central Ankara, in Dikmen and Öveçler, where they still actively work as waste workers.

Therefore, Kurdish migrants engaged in unrecognized recycling gradually increased in number, which has given these informal activities an ethnic character (see Njeru, 2006; Samson, 2010). These groups collected and sorted mainly paper, but also plastic, metal and glass, which was sold to factories all over Turkey. This led to a dualistic perception of waste with materials that carry economic value in informal buy-sell relations considered recyclables, and the remainder considered disposables. In short, accompanied by the valorization of waste, these political and economic forms of migration paved the way for the emergence of systematized, unrecognized recycling as a source of livelihood for these migrants (Gidwani, 2015; Gutberlet, 2016; Parizeau, 2015). The relation between the materiality and technologies of waste and the new urban poor acquired not only a methodical, but also an ethnic character.

Discussion: the entangled factors of transformation and the informalization of waste regimes in Ankara

This examination of the key moments and processes underlying the informalization of waste regimes in Ankara between the 1930s and the 2000s shows how a number of macro forces coalesced and provided the context for place-specific waste management practices. The political and economic changes (see Gidwani, 2015) and their unintended consequences gave rise to a number of material changes, namely in waste composition, waste technologies and the devalorization of land. These changes had direct ramifications on place-specific actor responses and played a major role in the production, transformation and informalization of waste regimes. The impact of broader political-economic decisions and their material consequences, and related institutional dynamics, is evident both in the shift from the organic regime to the regime of volume, and later, to the regime of value. The waste regimes in Ankara between the 1930s and the 1980s took shape under the restricted socio-technical capacities of the municipalities. During these periods, the state prioritized an industrial policy. Although this policy changed over the years (i.e., the shift from state-led industrialization to import substitution in the 1940s), this prioritization strictly restricted the budgets allocated to municipalities, and thus to urban infrastructure services. For this period, it is not possible to refer to informality as a dominant characteristic of the waste regimes.

Migration inflows, an unintended consequence of mechanization in the agriculture sector, resulted in demographic growth and an associated increase in the volumes of waste produced. The first formal responses from the municipality included formal open dumping practices at a designated site, far from the city and on unvalued, state-owned land. Secondly, a mass *gecekondu* development occurred near and on these municipal open dumping sites. Therefore, the major socio-technical decisions regarding waste focused on the spatial displacement of the open dumping sites. Accordingly, this research found the role of urbanization and the materiality of land as a widely neglected factor underlying waste regimes (for an exception, see Gidwani and Reddy, 2011). The early municipal open dumping practices on economically unvalued lands – including *gecekondu* developments on or near the open dumping sites – linked the socio-technical responses towards waste to the urban development in these areas. Increased amounts of waste along with valuable materials such as plastic, paper and coal resulted in the establishment of connections between the urban poor and waste through the emergence of unauthorized collection and reuse. Therefore, immanent relations between the materiality and technologies of waste, patterns of urbanization and the urban poor were established between the 1930s and the 2000s.

Waste management, specifically disposal, came in the form of a municipal response regarding its assigned duty to ensure urban sanitation and unfolded through ecologically deteriorating open dumping practices. The newly emerging forms of unrecognized recycling were also unmethodical and unsystematic. It could be argued that the composition of waste and the size of the industries that utilized recyclables (such as raw materials) did not constitute the necessary grounds for waste picking to become a systematic source of livelihood. Nonetheless, the context in which early forms of informality emerged was significantly transformed by the shift from import substitution to export-oriented industrialization in the 1980s. In the absence of statistical data from the period, the evidence from this research shows that this transformation in the production patterns changed the material composition of waste and increased the volumes of recycling materials such as rubber, textiles, glass, metals, different kinds of plastic, cardboard, paper, composite waste and bulky waste such as furniture, and electric and electronic waste. The increased volumes of recyclables in Ankara's waste resulted in a nuanced and dualistic perception of waste by the actors and redefined place-specific actor dynamics and social constructions of waste (Demaria and Schindler, 2016; Njeru, 2006). While materials that carried economic value for the socio-technical capacities of the waste sector were considered recyclables, any materials that could not be recycled within the current capacities were understood to be disposables and sent to open dumps.

The second determining factor is regulatory mechanisms and official institutional discourse. Gille's (2007) seminal work on the evolution of waste regimes in Bulgaria describes the dominant role of official discourses in determining waste regimes. Despite not being a former socialist country like Bulgaria, Turkey has a strong regulatory presence of state institutions. Legal regulations defining the scope and organizational terms of urban infrastructure services were extant. However, the changes in official discourse found no echo in practice since they did not define any clear financial tools or mechanisms for implementation. Until the

1990s, the Municipal Law (1930) and the Public Health Law (1930) remained the major institutional frameworks for regulating waste. The official discourse portrayed waste as an object/issue of municipal/public service, a matter of urban sanitation and cleanliness. The municipalities were the only authorized actors regarding local governance and the management of waste. These limitations manifested themselves with the emergence of municipal open dumping practices as the major socio-technical response towards waste. These practices took a path-dependent form and gave rise to future practices and decision-making processes. Later in the 1990s, the accelerating impact of international influences on the Turkish policy regimes became evident with the By-law on Solid Waste Control (1992) based on the EU waste framework directive. These regulatory changes were mainly related to the paradigm shift in the official discourse on waste, namely from a matter of sanitation to an environmental and economic concern (Güler et al., 2001).

Rather than the dominant official discourse, the institutional changes reassigning the responsibilities of waste treatment and disposal to metropolitan municipalities and the responsibility of waste collection to district municipalities have reshaped the waste regimes. The municipalities were given the power to privatize their waste-related responsibilities in public-private partnerships and by subcontracting. My findings describe the impact of this new ability to privatize. The unsuccessful municipal attempts to achieve technological change can be seen as having rendered the municipality a less powerful actor within the waste system. The capacity of the municipality to direct recyclable material flows was limited, and this opened up a potential economic space for the urban poor.

Thirdly, the labor dynamics of urbanization and the production of urban poverty played a major role in the informalization processes of Ankara's waste regimes. Gidwani's (2015) research on informal workers in India shows that Turkey is not a unique case of economic migration as a force in the production of informality in waste systems. The dissolution of agrarian economies in India also played a major role in triggering rural to urban migration (Sanyal, 2007). Campbell (2018) presents a similar case concerning the migrant waste workers of Thailand. However, the case of Ankara shows that changing political economies in the agricultural sector is not the only dynamic behind the production of migrant workers. Political migration was also important. The early 1990s witnessed a process that was defined by the head of the Recycling Laborers Association, Dinçer Mendillioğlu, as "the shift from forced migration to forced proletarianization" (2015). Accompanied by the valorization of waste, this political form of migration paved the way for the emergence of systematized, unrecognized recycling as a source of livelihood for these migrants. The relation between the materiality and technologies of waste and the new urban poor acquired not only a methodical, but also an ethnic character. The economic, but mainly political, migration of Kurdish groups to Ankara played a major role in the informalization of its waste regimes. What this research shows is that – unlike its common treatment in the literature – informality cannot be treated solely as a sectoral component or a given characteristic of waste regimes outside the context of the

Global South. The changes in these entangled factors of transformation jointly provide the context for the transformation in the understanding of waste from an externality to a resource (Demaria and Schindler, 2016) for authorized actors, along with it becoming a source of livelihood for the urban poor (Gutberlet, 2016; Parizeau, 2015).

Conclusion: approaching informality of waste regimes as a dynamic, social-historical process

This chapter focused on the informality of waste regimes (Gille, 2007) as a socio-historical process (Castells and Portes, 1989) that cuts across political ecology, urbanization and poverty. Its theoretical considerations were addressed by examining the changes in the relationship between poverty, urbanization and waste regimes in Ankara, Turkey, between the 1930s and the 2000s. The case allowed for an elaborate discussion of the links between the place-specific processes and policies of waste management in Ankara to the macro-economic, political and institutional transformations in Turkey. It showed how the social constructions of waste shifted from an externality to a resource (Demaria and Schindler, 2016), the value of which could not be realized under the limited capacities of municipalities. Coupled with the unintended consequences of macro-economic and political changes in Turkey, the limited socio-technical capacities of the formal actors led to the emergence of informal practices. These practices took on a methodical and systematic form after the mass political and economic migration of the ethnic Kurds. Thus, the systematization of informality is a consequence of changes in the systemic elements together with the contextual, place- and actor-specific motives, capacities, needs, claims and understandings of the actors. It creates socio-spatial, material and economic links between the urban poor, waste and urban growth in the course of rapid urbanization and population growth, which then reshapes waste regimes.

This research also points out the policy consequences of considering informality only as a given sectoral condition rather than a dynamic, social-historical process, which obscures the systemic dynamics that produce it. This leads to disregarding or not acknowledging the well-rooted existence of informal actors in waste systems, and public actors tend to look for solutions in techno-managerial interventions that fail to grasp the multifaceted nature of waste. These perspectives consider the integration of informal actors into formal systems as a matter of increasing efficiencies or effective system design (Wilson et al., 2006). Nonetheless, "solving waste problems can never become the exclusive domain of engineers" (Gille, 2007: 212). Given the limited political capacities of the informal actors, it also restricts the alternative to integration into the formal and recognized systems organized from the bottom and the betterment of everyday conditions only for small groups. Therefore, acknowledging the broader dynamics at play and redefining the scope of waste regimes' informality as more than solely an issue of effective waste management or the right to work could enable the engagement of a broader spectrum of societal actors, which could then create the power, pressure and financial mechanisms needed for a far-reaching change. The potentialities for and the limitations against the definition of such mechanisms stand as a major research agenda for future research on the informality of waste systems.

Acknowledgments

This research was partly accomplished as part of my doctoral studies on "Metabolic Urbanization of Waste in Ankara: A Governance Perspective" (Tuçaltan, 2017) with financial supports from German Research Foundation (DFG) research and training group "Topology of Technology" (GRK1343) at Darmstadt University of Technology, Germany, and Department of Human Geography and Planning at Utrecht University. I would like to express my gratitude to Prof. Jochen Monstadt and Prof. Roger Keil for their kind comments.

Notes

1 A series of interviews with long-term citizens of Ankara conducted in 2014 indicated that plastic shopping bags became a major component of urban waste in the 1960s.
2 A left-wing political organization active in many fields of urban political struggle such as the right to housing.
3 Turkish Statistical Institute. Prior to 2005, the institute was known as the State Institute of Statistics (*Devlet İstatistik Enstitüsü*; DİE).
4 The Scientific and Technological Research Council of Turkey.
5 According to a 1993 SIS survey, municipal solid waste in Turkey was 45.48% paper and cardboard, 8.62% metal, 18.46% glass, 13.19% plastic, 6.15% composite plastic (PET, PVC), 3.30% rubber and 4.80% textiles (TÜBİTAK, 2004).
6 The By-law was based on the European Commission's (1975) Waste Framework Directive (75/442/EEC).

Bibliography

Adama, O. (2014). Marginalisation and Integration within the Informal Urban Economy: The Case of Child Waste Pickers in Kaduna, Nigeria. *International Development Planning Review*, 36(2), pp. 155–180.
Aksoy, Ş. (1993). Turkish Experience with Privatisation: An Overview and Evaluation. *Turkish Public Administration Annual*, 17–19, pp. 39–54.
Alemu, K. (2017). Formal and Informal Actors in Addis Ababa's Solid Waste Management System. *IDS Bulletin*, 48(2), pp. 53–70.
Ankara Büyükşehir Belediyesi. (1993). *Ankara Programı 1993*, Ankara: Ankara Büyükşehir Belediyesi.
Ankara Valiliği Çevre ve Şehircilik İl Müdürlüğü. (2016). *Ankara İli 2015 Yılı Çevre Durum Raporu*. Ankara: Ankara Valiliği Çevre ve Şehircilik İl Müdürlüğü Çevresel Etki Değerlendirme Şube Müdürlüğü.
Arun, F. (1972). *Türkiye ve Dış Ülkelerde Çöp Konusu (Mevzuat ve Uygulamalarıyla)*. Ankara: Kardeş Matbaası.
Balaban, O. (2008). *Capital Accumulation, the State and the Production of Built Environment: The Case of Turkey*. Ph.D. Ankara: Middle East Technical University.

Belediye Kanunu (The Municipal Law). (1930). Resmi Gazete (Official Gazette), Tarih (Date): 14/04/1930. Sayı (Number): 1471.

Bjerkli, C. L. (2013). Governance on the Ground: A Study of Solid Waste Management in Addis Ababa, Ethiopia. *International Journal of Urban and Regional Research*, 37(4), pp. 1273–1287.

———. (2015). Power in Waste: Conflicting Agendas in Planning for Integrated Solid Waste Management in Addis Ababa, Ethiopia. *Norwegian Journal of Geography*, 69(1), pp. 18–27.

Boratav, K. (2015). *Türkiye İktisat Tarihi 1908–2009*. Istanbul: İmge Kitabevi.

Campbell, S. (2018). Migrant Waste Collectors in Thailand's Informal Economy. *European Journal of East Asian Studies*, 17(2), pp. 263–288.

Çamur, C. K. and Yenigül, S. B. (2009). The Rural-Urban Transformation Through Urban Sprawl: An Assessment of Ankara Metropolitan Area. In *The 4th International Conference of the International Forum on Urbanism (IFoU)*, 1st ed. Amsterdam/Delft, pp. 1045–1054.

Castells, M. and Portes, A. (1989). World Underneath: The Origins, Dynamics, and Effects of the Informal Economy. In A. Portes et al., eds, *The Informal Economy*, 1st ed. Baltimore, MD: Johns Hopkins University Press, pp. 11–37.

Cornea, N., Veron, R., and Zimmer, A. (2017). Clean City Politics: An Urban Political Ecology of Solid Waste in West Bengal, India. *Environment and Planning A*, 49(4), pp. 728–744.

Corwin, J. (2018). *Circuits of Capital: India's Electronic Waste in the Informal Global Economy*. Ph.D. Minneapolis: University of Minnesota.

Davis, M. (2006). *Planet of Slums*. London: Verso.

Davoudi, S. (2009). Scalar Tensions in the Governance of Waste: The Resilience of State Spatial Keynesianism. *Journal of Environmental Planning and Management*, 52, pp. 137–156.

Demaria, F. and Schindler, S. (2016). Contesting Urban Metabolism: Struggles over Waste-to-energy in Delhi, India. *Antipode*, 48(2), pp. 293–313.

Dilek, F. (2006). Mamak Düzensiz Depolama Alanı İçin Peyzaj Onarımının Önemi ve Gereği. *Tarım Bilimleri Dergisi*, 12(4), pp. 323–332.

Erberk, D., Arıkan, A., Öztürk, İ. K., and Tüzel, N. (2011). Turkish Packaging Industry Report. *Ambalaj Bülteni*, pp. 42–48.

Erman, T. (2004). Gecekondu Çalışmalarında 'Öteki' Olarak Gecekondulu Kurguları (Representations of the Gecekondu Inhabitant as 'Other' in Gecekondu Studies). *European Journal of Turkish Studies*. [online] Available at: http://ejts.revues.org/85 [accessed on May 2, 2017].

European Commission. (1975). *Council Directive 75/442/EEC of 15 July 1975 on Waste*. Brussels: The Council of the European Communities.

Fredericks, R. (2014). Vital Infrastructures of Trash in Dakar. *Comparative Studies of South Asia, Africa and the Middle East*, 34(3), pp. 532–548.

Gidwani, V. (2015). The Work of Waste: Inside India's Infra-Economy. *Transactions of the Institute of British Geographers*, 40(4), pp. 575–595.

Gidwani, V. and Reddy, R. N. (2011). The Afterlives of "Waste": Notes from India for a Minor History of Capitalist Surplus. *Antipode*, 43(5), pp. 1625–1658.

Gille, Z. (2007). *From the Cult of Waste to the Trash Heap of History: The Politics of Waste in Socialist and Postsocialist Hungary*. Bloomington: Indiana University Press.

———. (2010). Actor Networks, Modes of Production, and Waste Regimes: Reassembling the Macro-Social. *Environment and Planning A*, 42(5), pp. 1049–1064.

———. (2012). From Risk to Waste: Global Food Waste Regimes. *The Sociological Review Monograph Series*, 60(S2), pp. 27–46.

Güler, B. A., Atlı, A., Alıca, S., Güzel, G., Dalgıç, D., Göktürk, A., Özdemir, F., and Doğan, A. E. (2001). Politika. In B. A. Güler, ed., *Çöp Hizmetleri Yönetimi*, 1st ed. Ankara: TODAİE Yayını, pp. 10–39.

Gutberlet, J. (2008). *Recovering Resources – Recycling Citizenship: Urban Poverty Reduction in Latin America*. Burlington, VT: Ashgate.

———. (2012). Informal and Cooperative Recycling as a Poverty Eradication Strategy. *Geography Compass*, 6(1), pp. 19–34.

———. (2016). *Urban Recycling Cooperatives: Building Resilient Communities*. London: Routledge.

Gutberlet, J. and Jayme, B. D. O. (2010). The Story of My Face: How Environmental Stewards Perceive Stigmatization (Re) Produced by Discourse. *Sustainability*, 2(11), pp. 3339–3353.

İKMİB (İstanbul Kimyevi Maddeler ve Mamulleri İhracatçıları Birliği – Istanbul Chemicals and Chemical Products Exporters' Association). (2013). *Plastik Ambalaj Sektörü Gelecek Araştırması 2015–2023 Sonuç Raporu*. Antalya, Turkey.

Inverardi-Ferri, C. (2018). The Enclosure of "Waste Land": Rethinking Informality and Dispossession. *Transactions of the Institute of British Geographers*, 43, pp. 230–244.

Jaglin, S. (2014). Regulating Service Delivery in Southern Cities: Rethinking Urban Heterogeneity. In S. Parnell and S. Oldfield, eds, *The Routledge Handbook on Cities of the Global South*, 1st ed. Oxford: Routledge, pp. 434–447.

Katı Atıkların Kontrolü Yönetmeliği (By-law on Solid Waste). (1991). Resmi Gazete (Official Gazette), Tarih (Date): 14/03/1991. Sayı (Number): 20814.

Keskinok, Ç. (2006). *Kentleşme Siyasaları*. Ankara: Kaynak Yayınları.

Keyder, Ç. (1987). *State and Class in Turkey: A Study in Capitalist Development*. London: Verso.

Lawhon, M., Ernstson, H., and Silver, J. (2014). Provincializing Urban Political Ecology: Towards a Situated UPE Through African Urbanism. *Antipode*, 46, pp. 497–516.

Lawhon, M., Millington, N., and Stokes, K. (2018). A Labour Question for the 21st Century: Perpetuating the Work Ethic in the Absence of Jobs in South Africa's Waste Sector. *Journal of Southern African Studies*, 44(6), pp. 1115–1131.

Meagher, K. (1995). Crisis, Informalization and the Urban Informal Sector in Sub-Saharan Africa. *Development and Change*, 26(2), pp. 259–284.

Medina, M. (2007). *The World's Scavengers: Salvaging for Sustainable Consumption and Production*. Lanham: AltaMira Press.

———. (2011). The Informal Sector: A Driving Force for Recycling Management. In S. Spies, ed., *Recovering Resources, Creating Opportunities: Integrating the Informal Sector into Solid Waste Management*, 1st ed. Eschborn: Deutsche Gesellschaft für Internationale Zusammenarbeit (GIZ), pp. 9–11.

Mendillioğlu, D. (2015). Zorunlu göçten zorunlu işçileşmeye görünmez insanların hayata tutunmasının adı: Geri dönüşüm işçileri. *Halkçı*. [online] Available at: http://halkci.org/zorunlu-gocten-zorunlu-iscilesmeye-gorunmez-insanlarin-hayatatutunmasinin-adi-geri-donusum-iscileri/.

Millington, N. and Lawhon, M. (2018). Geographies of Waste: Conceptual Vectors from the Global South. *Progress in Human Geography*. doi: https://doi.org/10.1177/030913251 8799911.

The Ministry of Environment and Urbanization of the Republic of Turkey. (2016). *State of the Environment Report for Republic of Turkey*. Available at: https://webdosya.csb.gov.tr/db/ced/editordosya/tcdr_ing_2015.pdf [accessed on February 28, 2017].

Myers, G. (2005). *Disposable Cities: Garbage, Governance and Sustainable Development in Urban Africa*. London: Routledge.

Njeru, J. (2006). The Urban Political Ecology of Plastic Bag Waste Problem in Nairobi, Kenya. *Geoforum*, 37(6), pp. 1046–1058.

Nunn, N. and Gutberlet, J. (2013). Cooperative Recycling in São Paulo, Brazil: Towards an Emotional Consideration of Empowerment. *Area*, 45(4), pp. 452–458.

Nzeadibe, T. C. (2013). Informal Waste Management in Africa: Perspectives and Lessons from Nigerian Garbage Geographies. *Geography Compass*, 7(10), pp. 729–744.

O'Hare, P. (2017). *Recovering Requeche and Classifying Clasificadores: An Ethnography of Hygienic Enclosure and Montevideo's Waste Commons*. Ph.D. Cambridge: University of Cambridge.

Oteng-Ababio, M. (2018). Crossing Conceptual Boundaries: Re-Envisioning Coordination and Collaboration among Women for Sustainable Livelihoods in Ghana. *Local Environment*, 23(3), pp. 316–334.

Özbilgen, F. (1986). Büyük Kentler Büyük Dertler. *Cumhuriyet*, p. 6.

Özuğurlu, M. (2013). A real life "Grapes of Wrath". *Perspectives Turkey*, 6, pp. 31–34.

Parizeau, K. (2015). Urban Political Ecologies of Informal Recyclers' Health in Buenos Aires, Argentina. *Health & Place*, 33, pp. 67–74.

Samson, M. (2010). Producing Privatization: Rearticulating Race, Gender, Class and Space. *Antipode*, 42, pp. 404–432.

Sanyal, K. (2007). *Rethinking Capitalist Development: Primitive Accumulation, Governmentality and Post-Colonial Capitalism*. New Delhi: Routledge.

Schindler, S. and Kishore, B. (2015). Why Delhi Cannot Plan Its "New Towns": The Case of Solid Waste Management in Noida. *Geoforum*, 60, pp. 33–42.

Şengül, H. T. (2009). *Kentsel Çelişki ve Siyaset Kapitalist Kentleşme Süreçleri Üzerine Yazılar*. Ankara: İmge Kitabevi.

Şentürk, B. (2015). *Bu Çamuru Beraber Çiğnedik Bir Gecekondu Mahallesi Hikâyesi*. Ankara: İletişim Yayınları.

Simatele, D., Dlamini, S., and Kubanza, S. (2017). From Informality to Formality: Perspectives on the Challenges of Integrating Solid Waste Management into the Urban Development and Planning Policy in Johannesburg, South Africa. *Habitat International*, 63, pp. 122–130.

Sistem Planlama Müşavirlik Mühendislik ve Proje LTD. ŞTİ. (1992). T.C. Ankara Büyükşehir Belediyesi, Ankara Katı Atık Yönetimi Uygulama Projesi-Ankara Katı Atık Depolama Sahası Yer Seçimi Fizibilite Raporu Rapor No:1.

Tankut, G. (1990). *Bır Başkentin İmarı: Ankara, 1929–1939*. Ankara: ODTÜ.

tbmm.gov.tr. (1985). *The Grand National Assembly of Turkey, Official Minutes*. [online] Available at: www.tbmm.gov.tr/tutanaklar/TUTANAK/TBMM/d17/c019/b015/tbmm170190150370.pdf.

TÜBİTAK. (2004). Devlet İstatistik Enstitüsü ve Çevre İstatistikleri. *Vizyon 2023*. [online] Available at: www.tubitak.gov.tr/tubitak_content_files/vizyon2023/csk/EK-1A.pdf [accessed on December 16, 2016].

Tuçaltan, G. (2017). *Metabolic Urbanization of Waste in Ankara: A Governance Perspective*. Rotterdam: Eburon.

Umumi Hıfzıssıhha Kanunu (Public Health Law). (1930). Resmi Gazete (Official Gazette), Tarih (Date): 06/05/1930. Sayı (Number): 1489.

van Horen, B. (2004). Fragmented Coherence: Solid Waste Management in Colombo. *International Journal of Urban and Regional Research* 28(4) pp. 757–773.

Varley, A. (2013). Postcolonialising Informality? *Environment and Planning D: Society and Space*, 31(1), pp. 4–22.

Wilson, C., Velis, C., and Cheeseman, C. (2006). Role of Informal Sector Recycling in Waste Management in Developing Countries. *Habitat International*, 30(4), pp. 797–808.

Wittmer, J. and Parizeau, K. (2018). Informal Recyclers' Health Inequities in Vancouver, BC. *New Solutions: A Journal of Environmental and Occupational Health Policy*, 28(2), pp. 321–343.

Yeldan, E. (2006). Neo-Liberal Global Remedies: From Speculative-led Growth to IMF-led Crisis in Turkey. *Review of Radical Political Economics*, 38(2), pp. 193–213.

Yörür, N. (2010). 1990 Sonrası Türkiye'de Uygulanan Kırsal Alan ve Tarım Politikaları Üzerine Genel Bir Değerlendirme. *Planlama*, 1, pp. 3–19.

Young, O. R. (1982). *Resource Regimes: Natural Resources and Social Institutions*. Berkeley: University of California Press.

Part three

Movements on the edge

8 Contextualizing the rise of environmental movements in Turkey

Two instances of anti-gold mining resistance

Zehra Taşdemir Yaşın

In his study of the environmentalism of the poor, Rob Nixon (2011: 17) examines the global upsurge of indigenous resource rebellions especially in the neoliberal era by identifying "a clash of temporal perspectives between the short termers who arrive (with their official landscape maps) to extract, despoil, and depart and the long-termers who must live inside the ecological aftermath and must therefore weigh wealth differently in time scales". In similar vein, in this study, I explore the rise of environmental movements in Turkey as a clash between the process of transformation of nature into a field of resource extraction in the short term and the relationship between people and nature formed in the *longue durée*, through two historically prominent instances of anti-mining movements that emerged in the 1990s in Turkey, i.e., Bergama and Artvin-Cerattepe.

Since the 1990s, there has been a substantial rise in the number and spatial extent of environmental movements in Turkey. Conservation of green spaces, anti-mining, anti-reconstruction, anti-deforestation and anti-hydroelectric power plants have been some of the prominent focal points of these movements. While public opinion in Turkey has dominantly perceived these movements solely through the concerns of environmental protection, scholarly works have mostly focused either on the conservationist aspects of them or on the broader criticism that they direct at the dominant political economic system, i.e., neoliberalism. In order to bring ecological and political-economic dimensions of these movements into a relational analysis, this chapter underscores the historically specific socio-ecological context of the relationship between capitalist development and nature underlying the upsurge in environmental movements in Turkey.

In contextualizing the recent rise of environmental movements in Turkey historically, this chapter makes, at the most basic level, two interrelated propositions. Firstly, it proposes reconsidering the historical stance for environmental conservation as an internal and necessary constituent of a struggle for life and livelihood, which is a socio-ecological process. Secondly, it suggests reconceiving these movements as outcomes of geographically and historically specific relationships between capitalist development and socio-ecological change. Thought in this way, most of the environmental movements can be assessed as products of the deepening and expanding interaction of capitalist relations with nature through the formation of (extractive) resource frontiers and the ensuing socio-historical reconfiguration of human-nature interaction. Based on the examination of Bergama and Cerattepe struggles against gold (and copper) mining projects, I argue that they exemplify local articulations of a global environmental justice movement (Martinez-Alier, 2001, 2002; Martinez-Alier et al., 2016) against a geographically differentiated global process of socio-ecological transformation through the expansion of capitalist relations of production and extraction. The examination of these two local instances is demonstrative of how the language of resistance in these struggles has been gradually transformed to reflect this historical context of their emergence.

Joan Martinez-Alier (2012) observes through various environmental justice movements throughout the world that as the frontiers of resource extraction reach the remotest areas of the planet imposing the logic of the capitalist market, they are resisted by locally specific yet globally connected counter-movements asking for respect and recognition of their indigenous rights to sustain their lives and livelihoods based on mutual interaction with nature. Martinez-Alier and others (2016: 747) observe through the environmental justice atlas that:

> "there is a global movement for environmental justice, although almost all conflicts in the environmental justice atlas are local and they target specific local grievances. The movement is global because such local events belong to classes of conflicts that appear regularly elsewhere in the world (e.g. on open-cast copper mining, on oil palm plantations), or because they raise the conflict issue to a global level through movements' connections and networks and, by doing so, they actually create and operate at a global scale."

Despite their appearances in locally distinctive forms, the emerging environmental movements across the world have a common socio-historical denominator: they problematize the threat against the locally distinctive long-historical socio-ecological relations posed by the expansion of frontiers of resource extraction. The expansion of resource frontiers and their local environmental impacts have been studied through concepts such as "accumulation through dispossession" (Sneddon, 2007), "dispossession by accumulation" (Perreault, 2013) or "accumulation by environmental dispossession" (Latorre et al., 2015). These accounts extend or transcend David Harvey's (2003) conceptualization of "accumulation by dispossession" denoting ongoing processes of primitive accumulation that Marx talks about with respect to processes such as privatization of land and common/collective rights, expulsion of peasant populations and usurpation of commons that have resurfaced with neoliberalism. They argue that nature's materiality conditions processes of accumulation by dispossession (Sneddon, 2007), locate dispossession of livelihoods in the processes of accumulation of waste and expansion of the spatial footprint of extraction (Perreault, 2013), and specify the transformation of nature into a frontier and source of capital accumulation through modification of its biophysical characteristics and the shift of the environmental burdens of the extractive industries to these frontiers (Latorre et al., 2015: 66–67). I argue elsewhere (Yaşın, 2017) that beyond the recognition of the environmental impacts or burdens of the material production of extractive frontiers, what is of central concern is the socio-historical integration of nature into the world-historical relation of capital in these frontiers. This understanding recognizes the historical transformation of socio-ecological relations through the integration of nature into the processes of capital accumulation and revaluation of nature in contradiction to its locally distinctive long-historical valuation. This contradiction underlies, in this approach, the main source of environmental resistances.

In line with this approach, I explore Bergama and Artvin-Cerattepe anti-mining movements as instances in which to locate the local and global historical dimensions of the rise of the environmental movements in Turkey. Bergama movement represents "the first organized opposition to the operation of an MNC" following the adoption of neoliberal economic policies in Turkey (Özen and Özen, 2011). Artvin-Cerattepe movement represents one of the longest standing instances of environmental movements in Turkey since the 1990s, as its foundation was laid in 1995 by the establishment of *Yeşil Artvin Derneği* (*YAD*; Green Artvin Association) after the first drilling operations (Karahan, 2016). Both movements have struggled against the misrecognition of their social, cultural and material ties to land and ecology through the approval and development of mining projects in their vicinities, like their contemporaries taking place in other parts of the world (see Urkidi and Walter, 2011; Özkaynak and Rodríguez-Labajos, 2012). In this chapter, I examine the global and local socio-historical contexts through which Bergama's villages and the Cerattepe hill adjacent to Artvin were transformed into mining frontiers and the emergence and development of environmental justice movements against this transformation. I argue that although these movements were spawned in locally specific political ecologies of conflict, they were locally differentiated outcomes of a similar transformation of human-nature interaction, i.e., social abstraction of nature through capitalist accumulation processes and alienation of the human-nature relation. They reveal socio-historically contradictory forms of valuation of nature and their vocabularies of protest evolved to reflect this broader contradiction in the course of the struggle.

An outline of Bergama and Cerattepe anti-mining resistances

A thorough comprehension of the development of environmental resistance in local contexts depends upon an understanding of the geographically distinctive social, cultural and material interdependence with nature, which I refer as *human-nature unity*[1]. This unity, which is also underscored through a notion of symbiotic relationality between community and environment in the context of Bergama (Çoban, 2004), can take materially and culturally differential forms. In the Bergama mining frontier, it was based dominantly on rural-agricultural production relations, in which the long-historical and geographically distinctive material relation to land configured the socio-ecological and socio-cultural components of livelihood and living. In the Artvin-Cerattepe mining frontier, by contrast, it was based on ecological integration between wilderness, rural and urban "life spaces", in which nature conditioned urban life as a source of climate and water security and of topographical integrity and yielded ecologically based livelihoods such as forestry, apiculture and ecological tourism (TMMOB, 2014; Kalın, 2016; Karahan, 2016; *Evrensel*, August 14, 2017). Violating nature through extractivism misrecognizes, disintegrates and, thereby, violates this human-nature unity and also "the right to life" triggering a resistance movement (Martinez-Alier, 2002; Çoban, 2004). How the development of the mining sites reconfigures and dissolves this unity and how the state apparatus positions itself with respect to emerging conflicts between different value relations with nature lead to formation of historically new political frontiers, as happened through the development of the Bergama and Cerattepe environmental movements. In this respect, this section outlines the main socio-ecological and political-ecological constituents of the development of anti-mining movements in Bergama and Artvin-Cerattepe (see figure 8.1). The main socio-cultural and ecological impacts of the mines and the claims raised by each movement are briefly summarized in Table 8.1.

Bergama is covered by the most fertile lands in Anatolia located 118 km east of Izmir. It is also a vital touristic historical city hosting numerous historical and archaeological sites; 40 percent of the population of Bergama, which amounts to approximately 102,000, still consists of rural agricultural population. This ratio was 43 percent in 2007 and 51 percent in 2000 (tuik.gov.tr, 2018). In 2000, agricultural production was absorbing 70 percent of the labor force in Bergama. Cotton, tobacco and olives were the prominent agricultural products; while wheat, vegetable and fruit production, sheep and goat farming and dairy production constituted other agricultural livelihood activities of lesser importance. The agricultural production processes were predominantly based on seasonal cycles and unit of production was peasant families in the villages. Yet, they were integrated

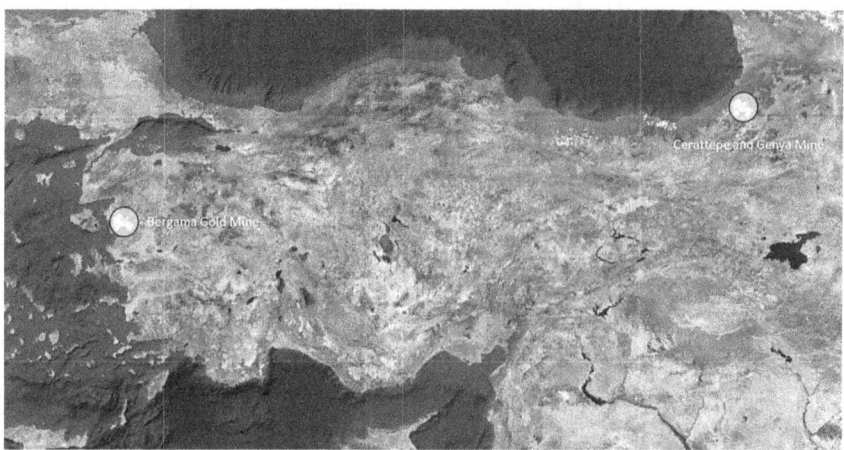

Figure 8.1 Bergama and Artvin (Cerattepe and Genya) Gold (and Copper) Mines[2]

in the local/national market economy and provided considerable sources of rural income (İZKA, 2013).

The decline in the agricultural population in Bergama in the last two decades can be explained by two mutually reinforcing processes: the formation of a mining frontier in Bergama and the integration of agricultural production in Turkey into the world market that gained momentum in the aftermath of Turkey's 2001 financial crisis. As underscored by several studies (Aydın, 2010; Keyder and Yenal, 2011; İslamoğlu, 2017; Adaman et al., 2018), the 2001 crisis paved the way for a neoliberal structural adjustment process in agriculture[3] that involved deregulation of agricultural production, privatization of regulatory agricultural institutions and liberalization of the agricultural economy. By reformulating cooperatives and instituting contract farming as the main link between local producers and transnational and national agribusiness companies, this neoliberal transition created a ripe plane for the substitution of traditional crops with high value added agricultural commodities and the expansion of global agribusiness capital in Turkey, shifting the market control from local producers and cooperatives to the MNCs (Aydın, 2010; İslamoğlu, 2017). The major outcomes of the global market integration of the agricultural economy of Turkey have been socio-ecological and economic reconfiguration of the rural space and de-agrarianization (Aydın, 2010; Keyder and Yenal, 2011; Adaman et al., 2018). Nonetheless, de-agrarianization has not translated into an equal decline in rural population especially in the coastal regions, where villages have gained in population. In other words, non-agricultural livelihood and employment sources have gained importance, indicating a process of "de-linking of rural livelihoods from land and farming" (Keyder and Yenal, 2011: 79). Mining has been one of such areas that has transformed the rural population into wage labor. Within this context, in the instance of Bergama, on one hand, the decrease in agricultural production in traditional crops including tobacco and cotton due to the neoliberal regulations that prevented state procurement, price supports and subsidies (Aydın, 2010; İslamoğlu, 2017) constituted the push factors for the expansion of wage labor. On the other hand, the formation of a gold mine or the rise of "extractivism" presented a strong pull factor away from agricultural production towards extraction (Adaman et al., 2018).

Bergama gold mine is located 12 km west of Bergama city center, in the vicinity of Ovacık, Çamköy, and Narlıca villages. After the Mining Law of 1985, an Australian mining company, Normandy, entered Turkey as one of the first foreign companies to have conducted gold exploration in Turkey. In partnership with the French company La Source and the Canadian company Inmet, Normandy established a multinational mining company named Eurogold in 1989 based in Turkey as required by Turkish law. After the initial exploration and drilling, Eurogold detected 24 tons of gold reserve to be extracted within a time span of 8 years and to be processed on site through the use of cyanide leaching technology and tailing ponds. It acquired land for its operations from local peasants living in Ovacık, Çamköy and Narlıca (*Milliyet*, October 21, 1994; Arsel, 2012: 124–125; Özen, 2009: 412). The mine was respectively acquired by Normandy from the Canadian and French partners in 1999, the US Newmont Company from Normandy in 2002, the Canadian Frontier Pacific Company in 2004 and the Turkish Koza Gold Company in 2005 (Özen, 2009).

According to the Environmental Protection Agency of the United States (EPA, 2009), the amount of toxic waste generated in hard-rock mining surpasses that of any other economic sector. Cyanide leaching in gold mining has been used since 1990s as a more efficient, feasible and profitable technology. It involves, however, higher environmental, social and public health risks and impacts (Urkidi and Walter, 2011). Bergama peasant resistance against the gold mine began through the initiative of local administrators and activists[4] who created awareness of the impact of cyanide leaching on public health and environment (Reinart, 2003: 55–57; *Milliyet*, June 6, 1991; *Milliyet*, April 9, 1992). The initial protests constructed a political frontier between Bergama community, who were directly exposed to the threats of the mining, and Eurogold, who was perceived as an "imperialist" power exploiting the local/national resources and environment. As such, the definition of the political opposition against the exploitation by a foreign company could socially and spatially expand the initially localized struggle by serving various urban-based and more educated groups as a political outlet for voicing their opposition to the neoliberal transformation of the national economy. While enabling the mobilization of more effective means of action including litigation processes and scientific reporting, these groups were influential in the extension of the political scope of the opposition with more emphasis on environmentalism and environmental protection (Özen, 2009).

With these diverse strategies and broadened political participation, the Bergama movement intensified and reached its apex in the second half of the 1990s, especially following a court decision in favor of the mining project in 1996 (Özen and Özen, 2009). According to Murat Arsel (2012: 124–126), in this period, it turned into "the epicenter of Turkey's most effective and visible environmental social movement against a multinational mining corporation" marking "a turning point in environmental politics in Turkey". It later became successful in winning the court cases that decided for the closure of the mine, thereby, preventing the operation of the Ovacık mine especially until 2002. There were three court orders, in 1997, 2002 and 2004 respectively, that cancelled the licenses of the company for the operation of mine (Özen and Özen, 2009, 2011; Reinart, 2003). Despite the legal decision for the suspension of the operations of the Company, however, the government publicly took side with the Company, manifesting it as an agent of economic development and did not implement the court orders. This reconfigured both the political frontiers and the scope of the political opposition. The Bergama movement turned into a nation-wide collective opposition not only against the Company, but also against the government and "the state-capital symbiosis". It demanded not only the closure of the mine, but also formed a collective identity demanding "rule of law" and "democracy" (Çoban, 2004; Özen, 2009). Thus, although the resistance initially challenged Eurogold, the official permits that it procured and the cyanide leaching method, it became politicized and challenged also the bureaucratic mechanisms, executive processes and the authoritarian rule especially in relation to environmental regulation and state-private capital relationship. A member of the *Çevre ve Ekoloji Hareketi Avukatları* (ÇEHAV; Environmental and Ecologist Movement Lawyers)[5] expressed this stance stating that the Bergama movement evolved into a "struggle of people, environment, and law against the state, government, multinational companies and their supporters" (Özen, 2009: 417).

As the anti-mining resistance expanded, the pro-mining politics and political actors, who involved, in addition to the company, state agencies, mainstream media and some academics, strengthened their position around the themes of "economic development" and "national interests" as well. They attempted to weaken the political credibility of the anti-mining resistance by publicly vocalizing the claim that the movement was pre-planned by a German organization for the purpose of preventing gold production in Turkey (Özen, 2009; Hablemitoğlu, 2003). This attempt to delegitimize the movement was instrumental in legitimizing more repressive interventions adopted by the government towards the anti-mining bloc as opposed to its more neutral stance vis-à-vis the company in the earlier stage of the struggle (Özen and Özen, 2009, 2011). The Council of Ministers' decision in 2002 that permitted the operation of the mine in contravention of the court decisions and enacted of a new mining law in 2004, further, engendered a social psychology of hopelessness in the movement. This psychology stiffened even more within the socio-economic context of high inflation and high unemployment created by the 2001 financial crisis in Turkey and by the process of structural transformation in agriculture, as briefly underscored above, that led to the devaluation of local agricultural products to the degree that they were inadequate for rural livelihoods. As a consequence, many peasants and young villagers, who participated in the anti-mining resistance for years, were compelled either to work in the mine or to migrate to the city center to find work (Reinart, 2003: 143–176; Adaman et al., 2018). A peasant from the village of Çamköy portrays this situation in 2001 as follows: "We have for years resisted with our hearts and souls for the future lives of our children. At present, there are people among us who want to work in the mine in this environment of unemployment and economic crisis. . . . The mining Company has a lot of money. They purchase our people and divide us making us enemies of ourselves" (Bektaş in Reinart, 2003: 143).

The evolution of the Cerattepe mining project and the struggle that it spurred pursued a divergent path from the Bergama instance. One of the factors underlying this distinctive trajectory is the social-ecology of Artvin, which has influenced bureaucratic mechanisms and judicial processes throughout the mining conflict. The social-ecology of Artvin mining site, which includes the Cerattepe area and the Genya Mountains, is a very unique product of ecological, topographical and social dynamics. The mine is located at a distance of 4 km southwest of the city center of Artvin and surrounded by small towns and villages in the south. As of 2015, Artvin had a population of around 168,000 of which 40 percent was rural population (tuik.gov.tr, 2018). Among the rural population, the main economic activities are composed of production of tea (10 percent of Turkey's whole production), hazelnut and fruits, animal husbandry, forestry and apiculture. In the urban areas, the composition of industrial production consists of production of food and forest products, mining and quarrying. Ecological and rural tourism has increasingly become an important economic activity and source of income. The distribution of the land is composed mainly of forests (57 percent), pastureland (nearly 15 percent) and agricultural land (around 9 percent) (doka.org.tr, 2018). The whole mining site, covering 31.8 hectares of land, is located within the forest area. It resides in a mountainous area which constitutes the highest points of the city, bearing high risk of landslides. The site contains water resources that provide the main sources of drinking water to Artvin. It also contains protected natural areas, including a national nature park and an ecological tourism site (TMMOB, 2014; Gümüş et al., 2013). The mining site resides within one of the 34 global biodiversity hotspots, i.e., the Caucasus, which has been endangered by an accelerating conservation crisis over the past decade (Şekercioğlu, 2011; Şekercioğlu et al., 2011a, 2011b). It is also located within one of the 200 priority ecological regions for global conservation, i.e., Caucasus-Anatolian-Hyrcanian Temperate Forests (Olson and Dinerstein, 2009).

The Cerattepe mining site not only contains gold, but also copper, silver and zinc in different proportions within an integrated structure of layers. The gold mining project in Artvin commenced with the entrance of the Canadian company Cominco in 1991. The mining site comprised then one area of 250 hectares in Cerattepe. The anti-mining movement in Artvin, known as the Cerattepe resistance, emerged in 1995. Since then, it has spawned various local organizations, such as the YAD, that have been deeply engaged in the course of the events, prepared reports, organized local participation and mobilization and followed the bureaucratic and judiciary mechanisms to ensure their sound working. Especially large-scale petitions succeeded the suspension of the operations of Cominco and the mine was taken over by Inmet Mining in 2002. Titled with a local name called Artvin Cooper Mining Company, Inmet Mining presented itself as a copper mining project to be operated in the Cerattepe area, omitting from the mining site the area belonging to the Hatila National Reserves. However, as it could not convince the people of Artvin and stop the resistance, which started to employ judiciary methods, it left the site in 2009. In 2012, state recommended the mining site through initiating a tender with a permit for extracting all four types of mines on the basis of a change in mining law in 2010 enabling a new license mechanism based on tender. As an outcome of this new legal-administrative structure, the size of the site expanded to an area of 4,406 hectares including a major portion of Genya Mountains area entailing within it the Kafkasör ecological tourism site located between the city of Artvin and the Cerattepe hills. The final winner of the tender was Cengiz Holding in 2013, entering the site with the company named Eti Bakır, which had been a state-owned enterprise before the large-scale privatization taking place after the 1980s (Doğru, 2017; Gümüş et al., 2013; Özgökçeler and Sevgi, 2016). Especially from 2014 onwards, following the recommencement of the operation of the mine by Eti Bakır, local mobilization intensified and the course of the struggle changed, turning into more visible and public forms of demonstrations and gaining national and international visibility and participation (Karahan, 2016; Erensü, 2014b; Hattam, 2015). Yet, especially from 2015 onwards, it has also witnessed fiercer response from the state, leading to its gradual weakening (Karahan, 2016, 2017).

In terms of the historical (re)configuration of political frontiers, the Artvin-Cerattepe and Bergama anti-mining movements represent distinctive temporalities of a historical continuum of the formation of an environmental justice movement in Turkey. In Cerattepe, the licensed mining projects developed by Cominco and Inmet could be removed through, respectively, petition and litigation-based opposition led by the YAD by the end of the 2000s (Karahan, 2016). The Cerattepe movement entered in its most intense stage after 2012, when the tender mechanism came into play in mining projects and Eti Bakır acquired the project. It intensified historically in an already politicized field that had been establishing at the national level in a form characterized by a resilient opposition between the state-capital relation that aimed at expanding the extractive or ecologically transformative material frontiers, on one hand, and the communities who want to protect their nature-based living ecosystems or urban life spaces (Kalın, 2016; Erensü, 2014b; Hattam, 2015), on the other. In Bergama's instance, the primary target of the opposition was a foreign "imperialist" company. When the more intense antagonism between the movement and the state, or "the state-capital symbiosis" (Çoban, 2004), developed, the movement also began to enter into a process of weakening in the context of the 2001 financial crisis and the process of structural transformation in agriculture. However, the intense stage of the Artvin-Cerattepe instance was built on an already established political antagonism between a critical ecological stance voiced by the movement and a materially expansionist capitalist stance supported by the state. As such, its weakening became possible from early 2015 onwards by state repression enabling the mining operations (Karahan, 2016).

How should we discern and interpret these rough narrations of the Bergama and Cerattepe mining areas, projects and movements? Although both of the mining projects, their impacts and the claims raised against them indicate local processes and politicization of local agitation, in both cases, the anti-mining resistance pulled nation-wide and even, to some degree, international support and were transformed into national level mobilizations in the course of the struggle. An important explanation for this is that they have incorporated diverse social groups to express their dissatisfaction with not only environment-related but also other prevailing social and economic policies as well as political and institutional arrangements (Özen, 2009). This approach renders some of the constitutive dynamics in the development and spread of these movements. Beyond this rendition, however, my purpose in this chapter is to situate these struggles in relation to the specific historical relationship between capitalist development and socio-ecological change, as a common denominator underlying the discontent of various social groups with the prevailing policies and processes of accumulation. I suggest comprehending these movements at a deeper and broader historical level of the relationality between capitalist development and socio-ecological transformation, i.e., transformation of the human-nature relationship. In this respect, the rest of the chapter explores from a world-historical perspective how the historically specific tension between environment and capitalist development, to rephrase Arsel (2012), has been developed in Turkey in relation to mining and how we can recognize and assess this tension as the core aspect of the development of the environmental justice movements studied here from within the movement itself.

Changing world mining economy and neoliberal transformation of mining in Turkey

Studies on the world mining economy and mining conflicts around the world demonstrate both a spatial expansion of *the commodity frontiers* (Moore, 2000) in mining in the Global South and a material expansion of extraction, with some distinguishing global trends taking place in the last two decades (Martinez-Alier, 2002; Tsing, 2003; Bridge, 2004; Krausmann et al., 2009; Muradian et al., 2012; Latorre et al., 2015; Watts, 2015). Firstly, the *social metabolism*[6] of the capitalist world economy has increased (Martinez-Alier, 2002; Krausmann et al., 2009; Muradian et al., 2012). While the world population increased by approximately 72 percent from 1970 to 2004, the rise in the global extraction of major metals and construction materials were 75 percent and 106 percent, respectively, in the same period (Özkaynak and Rodríguez-Labajos, 2012: 7). Secondly, the mining fields have been spatially relocated from the Global North to the Global South (Bridge, 2004). In other words, the increase in world's social metabolism has been possible by the spatial expansion of extractive frontiers and the commodification of nature in the (semi)periphery. Especially with the adoption of neoliberal economic policies and the consequent weakening of environmental and labor regulations, the (semi)peripheral countries have seen a dramatic increase in the mining activities and frontiers since the 1990s (see Latorre et al., 2015: 58–69). Those peripheral rural and peri-urban areas, such as Bergama and Artvin, which had been bypassed by the world mining industry, have been incorporated into the global mining economy as new extractive frontiers through widening and deepening processes of neoliberalization.

Besides the substitution of domestic material extraction in industrialized countries by the imports from the rest of the world, another important aspect of the global spatial relocation of the world mining industry has been the emergence of Asian economies as major raw material consumers. In the last two decades, China has replaced the US position to be the largest consumer and importer of "natural resources" from mainly Canada, Australia, Latin America and Africa. An implication of this transformation has been the de-linking of the emerging markets or traditional peripheries, including Turkey, from traditional economic cores in the global mining economy through the increased Asian demand for raw material imports (Muradian et al., 2012). In 1980s and 1990s, trade relations between the periphery and the core in the mining sector were producing conditions that enforced the adoption of neoliberal policies and practices in the

Table 8.1 Main features of Bergama and Cerattepe conflicts[7]

	Bergama	Cerattepe
Characteristics of location	Rural-agricultural production Historical tourism	Rural and urban Animal husbandry Forestry Environmentally protected sites National park reserves Ecological tourism
Population under direct impact	Rural- agricultural population (village peasants)	Mostly urban population of the Artvin city center and rural-pastoral and forester peasant population
Distance between the project and the affected community	600 meters (to Ovacık village)	Some villages are enclosed within the site; 4 km to Artvin city center
Main local organizations involved	Bergama Municipality Executive Council of Environment (informal)	Green Artvin Association Artvin Bar Association
Local involvement	High	High
Intensity of the conflict (rough periodization)	1992–1996: low-to-medium 1996–1998: high 1998–2002: medium 2002–today: low	1994–2011: low-to-medium 2012–2016: high 2016–today: medium-to-low
Means of resistance	Awareness campaigns Road blockages Civil disobedience Referendum Occupation of the mine Judicial activism Demonstrations and street protests Networking with the municipal organs Networking with national and international NGOs	Awareness campaigns Road blockages Petitions Judicial activism Demonstrations and street protests Development of local community organizations Networking with national and international NGOs Meetings with the members of the Government
Outcomes of resistance	Court decisions (success) Governmental actions (failure): legislative changes, commissioned reports, the Council of Ministers decisions Technological improvements by the company Compensation of the affected families by the state	Court decisions (success until the final decision) Governmental actions (failure): legislative changes, commissioned reports, changes in the mining law (introduction of tender system) Cases of custodies and police injuries (suppression)
Main impacts of the mine to environment	Pollution of the underground and drinking water Soil pollution Deforestation	Water and soil pollution Deforestation and erosion Damages to biodiversity Damages to the natural protection sites Increased risks of landslide/flood

(Continued)

Table 8.1 (Continued)

	Bergama	Cerattepe
Adverse health effects of the mine	Health risks due to cyanide use	Health risks due to cyanide use
		Exposure to dust and noise
		Dead livestock due to water and soil contamination
Socioeconomic impacts of the mine	Adverse impact on agricultural production due to water pollution	Adverse impact on cattle production
	Decrease in agricultural lands due to the sale to the mine companies	Tourism income losses
	Agricultural income losses	Outmigration
	Tourism income losses	
	Migration from the nearby villages	
Cultural impacts of the mine	Loss of livelihoods	Loss of livelihoods
	Damages to the indigenous peasant culture and environmental values	Damages to urban life space
		Damages to environmental values
	Damages to historical sites	Damages to the peasant culture
Main claims	Right to healthy living	Right to protect green ecosystem, ecological integrity and diversity
	Right to agricultural production and peasant livelihood	Right to nature and human-nature unity
	Right to nature and human-nature unity	Right to livelihood

extractive activities in the (semi)periphery while favoring Western transnational companies. However, the rise of China and other Asian economies as major consumers in the global mining industry complicated these relations, inducing also the replacement of the transnational corporations by their national substitutes in some of the major mining projects as was apparent in Turkey in the 2000s as well.

This world-historical transformation of the global mining industry, as a part of the capitalist world economy, has also reconfigured the relationship between state, capital and environment in Turkey in relation to the spatial and material expansion of mining frontiers. In the last two decades, Turkey's social metabolism has enlarged in terms of the levels of material extraction and consumption. In only the last decade, while the total value of mining activities in the economy increased sixfold, the share of the mining sector in the national income rose from 1.1 percent to 1.5 percent. There is also a similar increase in both mining exports and imports and in the share of foreign trade in mining in the entire foreign trade from the mid-1990s to the 2010s. More specifically, while the total volume of gold produced in 2003 was 5.39 tons, in 2014 it rose to 31.26 tons.

The total volume of copper production rose twofold from 2004 to 2014 reaching 6.4 million tons.[8] These increases in values and shares of mining have been made possible by the expansion of the social and material space of mining and incorporation of more "reserve zones" into the mining economy of Turkey in the last two decades, of which Bergama and Cerattepe are two instances. Such material expansion, in turn, has been regulated through neoliberalization of the mining industry and commodification of nature since the 1980s.

A glance at the history of the regulation of the mining economy in republican Turkey before the 1980s shows a cycle of liberalism and national statism. In the period from 1923 to the early 1930s, foreign capital was dominant in the mining sector through the grants of concession rights, which was a dynamic that had already settled in the Ottoman state in the nineteenth century. According to Çağlar Keyder (1981), even the most radical context of the 1923 Izmir Economic Congress, which constituted a symbolic base for the re-establishment of Turkish national economy, abstained from a perception of keeping the foreign capital out of the Turkish economy. Especially in the industrial areas that required modern technology and high capital investments, the foreign direct investment was seen as a substantial means in the development of modern industry. By 1929, mining constituted the second-largest sector of foreign capital investment that extracted natural resources for export. For instance, in 1925, Deutsche Bank obtained the right to exploit Ergani copper mines together with the right to construct a railway to the mine. In fact, the exploitation of mines, together with agricultural integration, constituted one of the major impetuses for the construction of railways and determination of their routes in the early Republican era, a trait that was again inherited from the late Ottoman era (Ibid.: 29–31, 56–88, 103).

In 1935, a process of nationalization of mines commenced with the establishment of the *Maden Tetkik ve Arama Genel Müdürlüğü* (MTA; General Directorate of Mineral Research and Exploration) and Etibank through which the state assumed the main responsibility to invest and operate in mining. By the 1940s, the licenses of the mines that had been operated by the foreign capital were bought and transferred to Etibank. However, in the 1950s, the mining sector was re-opened to private capital investments in the context of a broader economic liberalization program. The 1954 Mining Law No 6309 granted equal rights to explore and operate mines to public and private capital enterprises. Nonetheless, in the 1960s, despite the fact that the 1954 mining law was intact, the regulative and protective role of the state in the economy resurfaced and was strengthened with the new Constitution that reassigned all the rights pertaining to natural resources to the disposal of state. Through the Law on Mines to be operated by the state approved in 1978, most of the private sector investments in mining were ceded to the state by the 1980s. In this period, thus, state-owned enterprises became the only agents that could conduct the extraction of "natural resources" (Tamzok, 2005).

Nonetheless, the 1980s witnessed a major transformation in the exploration and extraction rights of mining, inducing the integration of the national mines into the global mining industry. In this neoliberalization phase, the relationship between mines and national development was construed via increasing exports and export revenues through empowerment of private capital. In 1983, a new law was enacted that approved the return of those mines that were confiscated from the private capital by the 1978 law to the previous private owners. The mining law of 1985 enabled the privatization of mining operations and encouraged the foreign corporations to invest in mining projects in Turkey. From the 1990s onwards, Etibank-owned mining enterprises were swiftly privatized. The adoption of the Multilateral Agreement on Investment (MAI) principles into the constitution in 1999 under the recommendation of the IMF further reinforced liberalization and deregulation of the mining industry creating a more attractive space for especially foreign direct investment. While 91 percent of all the mining investments were publicly owned in 1980, 82 percent of all the mining investments became private capital investments by 2002 (Arsel, 2012; Tamzok, 2005; Özen, 2009; Konak, 2008). As already stated above, the first foreign company investing in mining in Turkey was Eurogold with its gold explorations in Bergama. The number of foreign gold mining corporations investing in Turkey rose to 12 on the eve of the court decision for the closure of Bergama mine in the early 2000s, while nine of them left the country after the decision (Özen and Özen, 2009). As a response, an amendment to the mining law enacted in 2004, further, rendered it more attractive for private and foreign capital. The new law expedited the process of obtaining licenses for exploration[9] and reduced the corporate tax and licensing fees for landholdings (Hurley and Arı, 2011). It also opened up formerly protected areas including olive fields, forests, national parks, naturally protected sites and historic sites to extractive activities on the condition of a decree approved by the Council of Ministers. With this amendment, the number of the companies that obtained licenses for the exploration and extraction of gold rose to sixty-one by 2008. In other words, the statist logic that regulated the mining field in Turkey until the 1980s has completely turned into the neoliberal logic since especially 1985 (Özen and Özen, 2009). As a consequence, the private investments have transformed many reserved zones into extractive mining sites as commodity frontiers supplying the world's raw material demands.

We can differentiate the neoliberal structure that the Turkish political-economic regime has attained since 1980s in a more nuanced way if we treat it roughly under two distinct phases. The main pillars of the first phase taking place from the 1980s until the late 1990s were programs of foreign trade liberalization, capital account liberalization and privatization in the context of a broader model of export-oriented industrialization as stated above. World-historically this phase unfolded in the context of the global relocation of mines or extractive processes from the core economies to (semi)peripheral regions. State-capital linkages in this phase favored foreign trade companies, which were family-run holding companies based mostly in Istanbul, through export incentive mechanisms provided by the state apparatus or the political elite (Karadag, 2010; Buğra, 1994: 146–155). The second phase, taking place especially since the mid-2000s, entailed a stronger system of state regulation especially of the financial system and deeper political involvement in the processes of capital accumulation of the business elite, while it has continued to strengthen the liberal market economy and foreign trade networks. In Weber's terms, a new "politically-oriented mode of profit-making" has been formed leading to the emergence and consolidation of a new state-led and Anatolian-based capitalist class, some of whom had begun to rise in the 1990s as "Anatolian tigers" or "green capital", benefiting from state resources and favors through new legislative and administrative mechanisms and network relations developing along the cultural, religious and political identities and ties (Buğra and Savaşkan, 2012; Karadag, 2010; Öncü, 2014).

A symptomatic example of this symbiotic rearrangement of the relationship between state and capital that was established in the processes of accumulation has been the energy sector (Erensü, 2018). Erensü's analysis (2018) of nationally based reformulation of the energy strategy shows that the relatively late neoliberal transition and privatization of the energy infrastructure has been constitutive of the consolidation of a more authoritarian political hegemony together with increased intervention into the processes of neoliberal capital accumulation. A similar process has been pertinent in the reconfiguration of the mining industry as well. While from the 1980s to the early 2000s, the attraction of foreign direct investment in mining was an important priority, from the mid-2000s onwards national private capital investments have also been favored. The changing world-historical structure of the world mining industry, in which Asian economies have become the major consumers especially from the 2000s onwards, has created fertile and more flexible conditions for such a reconfiguration as the newly emerging business elite could establish new trade networks with the rising Asian economies. In the two instances of the Bergama and Cerattepe gold (and copper) mines, the processes of the acquisition or purchase of the mines by private national companies, i.e., Koza Gold and Eti Bakır (Cengiz Holding) in 2005 and 2013 respectively, constitute a prototype of the political-economic context of the variable relationship between the political authority and the reforming capitalist class. As the following section deals with, this relationship has mattered in the politicization of the environmental movements in Turkey with respect to the revaluation of nature within the context of capitalist incorporation of nature and dissolution of local socio-ecological relations.

Neoliberal state's revaluation of nature

The transition to neoliberal regulation of the mining industry and the transformation of the relationship between state and capital have borne major consequences at the social and material level for human interaction with nature and valuation of nature. In their examination of the changing state, society and nature relationship via the hydroelectric power plant projects (HEPPs) and anti-HEPP resistance in Turkey, Adaman et al. (2016) examine such consequences in the context of the regulation of water resources for energy generation that has been integrated within a neoliberal logic of state-capital relation. They specify a shift in the state, society and nature relation from the developmental state-dam

axis from the 1960s until the 1980s to private capital-HEPP axis in the last decades. While dams were symbols of a process of modernization that was promised
to the whole public and undertaken by the developmental state, they argue that
the HEPPs emerge as private capital investments that privatize the energy services for the public. In other words, the latter symbolize the seizure of the water,
which is a natural constituent of local socio-ecologies, for the benefit of just one
segment of society, i.e., the capitalist class. In this regard, as they contend, the
mechanisms of obtaining the consent of the local communities in the context of
private capital investments that directly impact these communities are not successful enough to convince the public. They damage the principles of reciprocity
and equal equidistance, which are central to establishing and preserving state
hegemony in Gramscian terms. The selectively favoring political ties between
the state apparatus and private capital in these investments further impair securing hegemony through consent (Ibid.). They are even less successful in the case
of mining investments, as both the negative transformative impacts of extraction, i.e., dispossessory and alienating pressures, and the erosion of reciprocity
and equality are more direct and visible. In this respect, the political-economic
context shapes the evolution and nature of the anti-mining environmental struggles internally.

The neoliberal shift in the regulation of the mining industry has represented
simultaneously a shift in the regulation of human-nature interactions and valuation of nature. Like many other examples in the (semi)peripheral world, in Turkey,
the neoliberal state was also reformed as an "environmental state" that "sustains
the ecological basis of production" (Sneddon and Fox, 2012) and capitalist accumulation by regulating nature and access of capital to nature. In other words, the
neoliberal state enabled the integration of reserved zones of nature into capitalist
value relations, its revaluation in the context of capitalist relations of extraction
in a commoditized form and, thereby, its alienation from existent socio-ecological
unities between local people and nature. Environmental regulation has been one
area for the state to sustain the alliances with the political authority through the
selective distribution of economic benefits, as discussed above.[10] To put it differently, the neoliberal-environmental state apparatus has worked as a catalyzer
for the processes of dispossession of people from human-nature unities that are
sources of livelihood and revaluation of nature in an abstracted and socialized
form in the context of the materiality of capitalist extraction and production
processes. In assuming this role, the state has mediated between private capital
and nature rather as an authoritarian medium enabling the access of the capital
to the reserved zones of nature at the expense of local communities. In this way,
environmental state politicizes the processes of transformation of peripheral livelihood ecologies into capitalist extraction sites.

According to Adaman et al. (2016), the neoliberal transition in the mining
sector is still embedded in the dominant modernization paradigm underlying
the national development project that has been one of the main pillars of the
raison d'être of the Turkish state. In this paradigm, the idea of progress is conceived mainly via economic growth; while political, social and environmental

transformation induced by the catalysts of economic growth have been relegated to a status of unavoidable externalities of growth. However, what changes under the neoliberal regime is the source of the economic growth, i.e., private capital investments. Through the environmental conflicts that have emerged and spread in the neoliberal era, the bureaucratic elite has widely adopted the language of national development and defined "natural resources" as assets to be exploited as major sources of foreign exchange earnings, state revenues, economic growth and, consequently, national interests. Such "national macroeconomic accounting" does not recognize, however, the costs of displacement and dispossession of communities in the extractive commodity frontiers through the destruction of their socio-ecological sources of livelihood. Neither does it recognize the destruction of environmental patterns and processes such as biodiversity and climate (Martinez-Alier, 2012; Turhan, 2017).

Therefore, this developmentalist language has at once become less concrete and less ambivalent in terms of its legitimizing and hegemonic function in the context of the neoliberal regulation of mining, as the economic benefits have been unevenly distributed. Bergama and Cerattepe struggles are examples that explicitly unveil this. The shift in the valuation of nature by the state apparatus from nature as a public resource that can be utilized through the medium of technology for public welfare to nature as "resource subject to commodification and privatization" (Sneddon, 2007: 186) has been very evident and frequently materialized through numerous projects that spurred the emergence and development of various environmental justice movements with alternate forms of valuations. The non-commensurability and clash of "languages of valuation", as Martinez-Alier (2001, 2009) puts it, between nature-based rights and livelihoods on one side and economic gains embedded in capitalist accumulation on the other has led the indigenous communities to resist in order to gain recognition of their own value standards with the use of various vocabularies and strategies of protest. In the following section, I focus on these indigenous forms of valuation of nature and how their contestations with the state's and capital's valuation of nature shaped the vocabularies of protest in the context of the Bergama and Artvin-Cerattepe movements.

Local valuation of nature and historical evolution of the vocabularies of protest

I argue in this section that Bergama and Cerattepe anti-mining resistances exemplify what Martinez Alier (2002) calls environmental justice movements against capitalist valuation of nature. In other words, they are instances of a broader world-historical pattern of conflict between two incommensurable socio-historical forms of value, as framed by Martinez-Alier, in their particular local appearances based on their specific geographical and ecological characteristics. Mining projects do not only violate the environment independent of the livelihoods of people. They transform the human-nature interaction and the socio-cultural relations based on this interaction. Thereby, through environment, they violate the whole socio-ecological unity. An examination of the historical evolution of the language of resistance of the Bergama and Cerattepe movements renders that they share with their global contemporaries the common denominator of opposition against capitalist integration of the local socio-ecological relations at their core. In demonstrating this, I explore how the vocabulary of protest in the Bergama and Cerattepe movements developed to reflect the inner tension between indigenous valuation of nature by the local communities and capitalist valuation by the private capital and state. Revealing this tension demonstrates how these two movements, although being locally embedded, are integral parts of a global environmental justice movement.

We can characterize both the Bergama and Cerattepe movements as "corridors of resistance" against the transformation of their social and physical ecosystems into extractive frontiers or "mining districts" (Özkaynak and Rodríguez-Labajos, 2012). As discussed above, these corridors are products of the processes of the world-historical expansion of extractive frontiers of capital accumulation, the ensuing integration of reserved nature into the relation of capital and national regulation of this integration through a neoliberal accumulation regime. They also reflect the peculiar and diverse local ecological, social and cultural settings, out of which they emerge, and a resistance against the homogenization of this peculiarity and diversity in the formation of a social-ecology of gold (and copper) mining with its own material and social aspects and spatiality. In this respect, I argue that while differences in the vocabularies of protest used in Bergama and Cerattepe movements reflect their local socio-ecological peculiarities or diversity, the common grounds underlying them reflect a stance against the patterns of homogenization through the formation of capitalist mining districts. Moreover, I argue that the historical development of these two movements show a dynamic temporal trajectory from voicing the local ecological peculiarities in relation to gold mining in the beginnings of the resistance to voicing the core of the historical transformation taking place in both locations, i.e., the transformation of the historical value of nature, the separation of human-nature relation and the consequent transformation of livelihoods, in the later stages.

In *Varieties of Environmentalism*, Ramachandra Guha and Martinez-Alier offer the concept of "vocabulary of protest" as a substitute to Charles Tilly's concept of "repertoire of contention." Accordingly, the latter concept is an instrumental one that implies the choices of the social protesters from a broader set in defending their economic and political interests most effectively. They argue, however, that the purpose of the social protest is not only to defend their interests, but also to articulate a judgment or critique on the prevailing arrangements. In other words, the protesters convey "both statements of purpose and of belief" (Guha and Martinez-Alier, 1997: 13). In concurrence with them, this section examines the changing vocabularies of protest employed in Bergama and Cerattepe movements in order to inquire primarily about their political critique or beliefs, in addition to their purpose or interests, and the historical evolution of this critique. These vocabularies became also means to (re)build and convey the emergent political identities.

In the Bergama movement, the vocabulary of protest evolved by gradually incorporating a political-ecological critique of the disruption of human-nature unity. While it presented mainly an anti-cyanide critique in the beginning, it later incorporated first a critique of the economic impact of gold mining on livelihoods and then a broader yet deeper critique of dispossession and alienation. The movement initially employed the language of preventing the cyanide leaching technology used in gold mining as a critique of the public health consequences of this method. The risk perception with regard to the cyanide leaching method constituted the major stimulus for the initial environmental mobilization (Orhan, 2006). The vocabulary of protest, developed especially in the early to mid-1990s, articulated a stance that was not against gold mining in itself but against the threat of cyanide use on their health and life. "No to Cyanide", "We don't want to die", "Yes to life, no to poisonous gold", "Human life is more valuable than gold", "We don't want gold, but life" were the dominant manifestations of this language in this period (*Milliyet*, April 9, 1992; *Milliyet*, June 20, 1992; *Milliyet*, July 15, 1992; *Milliyet*, July 18, 1992; *Milliyet*, July 21, 1992; *Miliyet*, August 4, 1992; *Milliyet*, October 17, 1994). The organization of awareness campaigns by the mayor and the local activists that informed the public of the health consequences of the cyanide technology was effective in the formation of this vocabulary (Reinart, 2003). Yet, while the anti-cyanide stance was developed in the context of the initial exploration and drilling stages of the operations of Eurogold, it developed and internalized an environmentalist critique of the destruction of the local ecology and a broader language of livelihood, as the spatial construction in the mine site proceeded and extractive processes started.

As the mining site began to gain an initial spatial and material existence with the cutting of trees, acquisition of lands and demarcation of the extraction site with fences, the reality of environmental degradation began to manifest as well (Reinart, 2003: 35–54). For peasants, this was not a violation of their "outer space" to be protected but of their long-historical connection with nature, which was an intrinsic part of their life (Uncu, 2012). The statements of a 53-year-old peasant from a village of Bergama as early as 1994 illustrates the indigenous value attributed to human-nature interaction: "I was born and grew up in the village of Ovacık. I finished the school here. We did not cut even a branch of those pine trees [which were cut by the Company], although we had times that we were cold and frozen in the winters. We had times that we ran out of money. We did not cut. We also did not let anyone to cut them. We always watched for those pine trees. Pine tree means life, it provides air. They [the miners working for the Company] cut thousands of them" (Umaç in Reinart, 2003: 45). According to a local politician in Bergama, the cutting of the trees ignited peasants to involve more actively in the movement and to participate in the decision-making processes, which were rather initiated by local politicians and activists before. Thereby, the second and more intense phase of the movement commenced (Engel in Reinart, 2003: 51).

In this second phase, the ecological critique was united with the livelihood concerns that reflected the negative consequences of the gold mine for the agricultural production in the region. As the soil and water pollution created by the waste of extraction (at the test stage) impacted the sales of the agricultural products, the vocabulary of the peasant protest incorporated an economic critique of gold mining on their livelihoods. Through the protests taking place in 1996, the peasants were using a new language reflecting their own local peculiarities and socio-ecologically embedded values: "Bergama's real gold is its cotton and tobacco" (Konyar in Reinart, 2003: 61). The villagers realized that the extraction processes of gold threaten their interdependence with nature or, as Çoban (2004) puts it, the community-environment symbiosis as the source of livelihood. A peasant woman leader reflected this at a local meeting noting that "thousands of people rely on this land that they are going to destroy, and you can see that human beings are somehow rooted in 'the land like plants'" (cited in Çoban, 2004: 443). Therefore, for peasants opposing the mine and protecting their ecology meant protecting their livelihoods, homes and lands. In interviews conducted in the region in 1997, one peasant from Çamköy stated, for instance: "We are not poor. God has given everything to this region. There is cotton, olives, tobacco, husbandry" (Milliyet, January 14, 1997). Other peasants expressed their concerns by stating: "If the mine starts operation completely, neither cotton, nor tobacco nor olives will have any worth. Those experts who buy our tobacco said that they would not come next year. What are we going to produce and sell then, what are we going to eat?" (Milliyet, May 5, 1997). Their articulations pointed out that they were getting poorer and not richer with the extraction of gold (Milliyet, July 27, 1997).

In the 2000s, the realization of the concrete impact of the gold mine on a larger socio-ecological, material and spatial scale altered the vocabulary of protest further, transforming it into a general critique of dispossession, displacement and the delinking of the human-nature relation. Following obtaining a permit to shift from test production to actual production with a Council of Ministries decision in 2002, the company begun to expand the spatial reach of the mining site by purchasing more lands from the nearby villages, especially from Narlıca. Moreover, as the mining and the 2001 financial crisis in Turkey damaged the agricultural sources of income of the peasants, working in the mine or migrating to the city centers as a source of income became a necessity, especially for younger people in these villages. In other words, the gold mine began to impose the formation of a social-ecology of mining. While this changing context weakened the resistance, though in a more silent way, the vocabulary of the protest turned into a resentment of the destruction of the culture of solidarity in the villages, of the loss of the livelihood, of the separation of the people from land and, thereof, alienation of the human-nature relation. Therefore, the demands for saving the livelihood were united with the demands for saving the human-nature unity as the source of their social, ecological and cultural values. One peasant expressed this in the following expressions: "What do we eat if we migrate to the city? What happens to us? We have never lived in a city. We have lived in the village. . . . In the city, we will be nothing, we will perish; because we know only soil, we don't know anything else" (Milliyet, October 15, 2003; Reinart, 2003: 147–164).

Therefore, many peasants of Bergama underwent a process of dispossession of their access to nature, clean soil, water, their sources of agricultural income and socio-cultural bonds of solidarity through the spatial and material "footprint" of the gold mine and the accumulation of contamination (Perreault, 2013). What happened in the Bergama mining frontier was a gradual process of incorporation of nature into the materiality of the mine, of its delinking from the indigenous forms of human-nature unity and its transformation into a "natural resource" for the production of gold as a commodity exchanged in the capitalist world market (Nixon, 2011). This transformation in Bergama constitutes a locally peculiar instance of a world-historical process of capitalist valuation of nature or integration of nature into capitalist social relations (Yaşın, 2017). The Bergama movement in its mature form, thus, represents the contradiction between two historically distinctive socio-ecological relations of value.

The Cerattepe movement is another instance of this contradiction. The historical evolution of the vocabulary of protest in the Cerattepe resistance can be understood in two stages. The first stage began with Cominco's initiation of the mine in 1994 with the increasing visibility of its impact on the livestock and deforestation and it continued until the departure of the second multinational company Inmet from the mine in 2009. At this stage, the struggle focused on the local level of organization, developing expertise on the impact of mining on the region and creating awareness in Artvin. Having experienced what had happened in the Bergama movement and from the processes of social and environmental transformation in the Murgul and Çayeli copper mines in other parts of Artvin, the language of resistance could incorporate an environmental and political ecological stance against the mine and mining policies based on its specific impact on Artvin's peculiar environment and biodiversity. "Environment is not fashion, but life style", "Environment is a human right", "Don't let mine, don't wither Artvin's green" constituted the main language through which the resistance brought an environmental critique to state policies related to mining and environmental regulation. A petition campaign titled "Don't touch Artvin and people from Artvin" organized in 2003 was using "Artvin" to connote environment and defining those who tried to enable the opening of the mine as "the enemies of the environment" (Karadeniz and Kalın, 2016).

The Cerattepe movement or Artvin green movement entered into a more intense stage after 2011, as the Ministry of Energy and Natural Resources declared that they would initiate a tender for 1,343 mines including the Cerattepe and Genya gold and copper sites. Thereafter, the vocabulary of the protest evolved into a more relational and encompassing stance criticizing how the people of Artvin were forced to abandon their long-historical past, culture, lands and memories in order to enable their replacement by a capitalist ecosystem of mining (Bianet, February 17, 2012). The 2013 "No to Mine" protests started to mobilize the people from not only Artvin but from all over Turkey with a statement summarizing the condition of Artvin's people: "We are forced to choose between either leaving Artvin or resistance". The protests integrated a critique of the transformation of the human-nature relation and destruction of human-nature unity, emphasizing that conceding the mining projects to Cerattepe and Genya meant the extinguishing not only of the environment but also of the people, history and culture (*Milliyet*, April 7, 2013; *Milliyet*, August 7, 2013). In the protests taking place in 2016, the language evolved from "We are protecting green Artvin and the right to environment" into "We are protecting the right to live", indicating a perception of environment as *sine qua non* of living in Artvin. A 92-year-old woman from Artvin epitomized the indispensable relationality between ecology and human life that the resistance internalized in its vocabulary in the course of the long struggle in the following words: "I was born here in Artvin. I have lived Artvin with its environment. I want my grandchildren to live Artvin in that way too. Money comes and goes, gold comes and goes; but life remains" (*Cumhuriyet*, March 17, 2016).

The comparison of Bergama and Cerattepe anti-mining movements presents valuable insights on the global and local dynamics of environmental justice movements and on the nature of socio-ecological transformation within a context of the integration of peripheral nature into the capitalist accumulation processes. On one side, both movements exemplify a world historical pattern. In both instances, the vocabularies of protest gradually evolved to the point that reflects a broader critique of the alienation of human-nature relation and of the gradual replacement of the long-historical and geographically distinctive indigenous social-ecologies and valuations of nature by the spatially homogenous and alienating social-ecology of capitalist extraction. This pattern, thus, entails the development of environmental justice movements within a context of contradictory historical forms of valuation of nature. Martinez-Alier specifies this as "incommensurability in valuation." He argues that:

> whenever there are unresolved ecological conflicts, there is likely to be not only a discrepancy but incommensurability in valuation. . . . To see in statements about human rights, indigenous rights, sacredness, culture, livelihood, intrinsic natural values, a priori refusal of the techniques of economic valuation in actual or fictitious markets, indicates a failure to grasp the existence of value pluralism.
>
> (2001: 167)

To further his point, we can argue that the incommensurability between local-indigenous and global-capitalist forms of valuations is a product of contemporaneous yet historically different and contradictory value relations. Both Bergama's peasants and Artvin's people value nature in the context of their historically specific social-ecology of livelihood that is based on a mutual relation with nature, which is a constitutive part not only of production relations but also of culture, identity and community life (Avcı et al., 2010). However, the valuation of nature in the context of social-ecology of resource extraction presumes nature as an already "natural resource" and a "commodity" that can be integrated into the wheels of capitalist exchange, abstracting it from the existent human-nature unity and re-socializing it in the context of the capitalist world economy. This incommensurability between socio-ecologically distinctive value relations with nature constitutes the world-historical temporality of the rise of environmental justice movements.

On the other side, the comparison of the Bergama and Cerattepe movements is revealing in understanding the distinctiveness and diversity of the local context in which global patterns of environmental resistance develop. This complexity is firstly a product of the geographically differentiated "materialities of nature" (Sneddon, 2007) and of human-nature unity, which differentially shape the material and spatial formation of the extraction sites and the relationship of the communities with nature. The specific composition of the social and political constituents of "the community-based resistance movements" forms, in turn, as an outcome of the relationship between the community and the material environment (Çoban, 2004: 440) and how it is reconfigured through the extractive processes. The extraction site in Bergama was formed within a rural ecosystem in which nature and land were the main sources of their livelihoods, (re)production, homes, water and the materiality of socio-cultural linkages. As such, throughout the Bergama resistance, gold was compared with land with a claim that "our land is very fruitful and more valuable than gold" (cited in Çoban, 2004: 443). In relation to this, the Bergama resistance was primarily developed as a "peasant movement". Yet, the "peasant identity" was transformed and gained a complex political-ecological form entailing values on environmentalism, human rights, justice and democracy. In the long-standing process of resistance, the relationality of these diverse values engendered a critical and unified political ecological stance that perceived the disruption of the ecological habitat through extraction and extractivism as a threat not only to the social lives and livelihoods but also to the human-nature unity and the right to live in an ecologically integrated way (Uncu, 2012).

The extraction site in Cerattepe hill borders the city of Artvin on its southern outskirts as well as a few nearby villages. The rich nature of the region endangered by mining is the central component of an integrated rural-urban ecosystem of Artvin and the source of clean air, clean water and tourism-based livelihood activities for the urban life in the city of Artvin.[11] The mine threatens urban life in Artvin as a whole (Doğru, 2017). In an expert report, conducted in 2014 in the context of a lawsuit process filed by the protestors, it was emphasized that a mining operation would completely end the city of Artvin as a "life space" (Kalın, 2016). As such, one of the central expressions of the Artvin-Cerattepe resistance was that "the surface of Artvin is more valuable than its underneath". Although the peasants living in the villages surrounding the extraction site and the rural forms of protests, such as watching, have accompanied the resistance, in its most intense stages, the Artvin-Cerattepe movement was dominantly identified and politicized as an urban life space struggle (*Posta*, February 23, 2016). The idea of *life space* internalized in itself a locally-geographically specific form of human-nature integrity. Therefore, the local-geographical distinctiveness of the materiality of the ecologies of Bergama and Artvin demonstrate the diverse appearances and experiences of the transformations of human-nature relation and diverse sources of the emergence and development of environmental or socio-ecological resistance. Socio-ecologically distinctive local articulations of the contradictions between two historically distinctive value relations and valuations, in turn, give rise to the spatially diverse yet temporally, or world-historically, "identical" or equivalent environmental justice movements.

This equivalence is most notable in the solidarity among spatially diverse localities or communities resisting against the enforced transformations of their natural ecosystems and in the social and spatial expansion of environmental justice movements. For instance, in 2014, as a consequence of the intense opposition in Artvin, Eti Bakır declared that they were going to carry their planned gold processing operations with the cyanide leaching method to their cooper extraction site in Murgul, a nearby mining town, which had been contaminated by cooper extraction for decades. The response of the predominantly laboring population of Murgul to this was a collective opposition in the form of an "ecology-labor coalition" including a labor strike of over 6,000 workers (Erensü, 2014a). The Artvin community also supported this opposition by participating in the local demonstrations organized through a newly initiated Murgul No to Cyanide Platform (Erensü, 2014b; Hattam, 2015). Similarly, despite its development at a communal and local level, the Bergama movement did not show a parochial attitude and gave support to other movements. For instance, they established connections with the mobilization against Ilısu Dam and gold exploration in Southeastern Turkey. Or, they participated in anti-nuclear demonstrations against the plan of constructing a nuclear power plant in Akkuyu, Mersin (Uncu, 2012: 169–170). In both instances of Artvin and Bergama, the communities could shift the local scale of their movement to national and even global scales, because of the fact that they share a common historical ground with various environmental movements nationally and globally: they are opposing the integration of their socio-ecologies in capitalist accumulation processes and the imposed transformation of their relationship to nature.

Conclusion

The examination of Bergama and Cerattepe anti-mining resistances as prominent instances of the environmental movements that emerged in the last few decades suggests that we cannot fully grasp the historical nature of the rise of environmentally driven social movements in Turkey through either mere focus on ecological conservation or opposition to the existent national political-economic regime. What is more deeply at stake is the historical transformation of human-nature relation and dissolution of locally specific human-nature bonds through the global expansion of extractive or nature-based commodity frontiers. Conservational concerns are influential in the emergence of initial resistance and the broader political-economic discontent is influential in the spatial expansion of the movements. Yet, as the above instances suggest, the main motive that sustains the development of long-standing environmental movements and the solidarity among locally developed mobilizations is the gradual disturbance of the socio-culturally embedded human-nature unity, the dissolution of the material, social and cultural ties to nature and the consequent alienation of human-nature relation. Therefore, conservation and political economic discontent can emerge as two immediate and apparent faces of a more profound socio-ecological disorder formed through a world-historical process of capitalist incorporation of peripheral reserved zones of nature regulated through culturally differentiated forms of neoliberal regimes of accumulation at the national levels.

In this respect, the rise of environmental movements in Turkey can be better understood, if they are specified as parts of a global environmental justice movement that Martinez-Alier points out. Although this global movement is differentiated by spatially based local characteristics of ecology and geography, the materiality of commodity production and the socio-political nature of the state-society-nature relation, its main source independent of local particularities is the historical tension between the long historical human-nature unity and the social abstraction, revaluation and integration of nature in capitalist relations. This examination of Bergama and Cerattepe movements has shown how we can reconsider and situate the recent environmental movements in Turkey as a part of this global environmental justice movement.

Acknowledgment

I would like to thank Dale Tomich and Philip McMichael, the two anonymous referees and the editors for their elaborate and valuable comments, corrections and suggestions.

Notes

1 This notion goes back to Marx's ecological critique of capitalist relations. He uses the concept of *human-nature metabolism* to denote the metabolic interaction between society and nature as a basis to disclose the separation of this interaction through capitalist relations of production as a historic process (Marx, 1973: 489; Schmidt, 1965). I use the concept of *human-nature unity* to refer in addition to such material metabolic interaction the socio-cultural relations embedded in this metabolism as well.

2 Retrieved from Environmental Justice Atlas. https://ejatlas.org/country/turkey.

3 This process was enforced and implemented by the World Bank, IMF and, later, EU through a series of stabilization and reform programs and legislative changes including most significantly the Agricultural Reform and Implementation Project (ARIP) introduced in 2001, the agricultural reform process along the EU's Common Agricultural Policy following the commencement of the EU accession negotiations in 2004, 2001 Tobacco Law, 2006 Agrarian Law and 2006 Seeds Law (see Aydın, 2010; İslamoğlu, 2017).

4 The mayor of Bergama, Sefa Taşkın, and a local activist and politician, Oktay Konyar, were the prominent figures canalizing the mobilization and organizing peasants.

5 It is a consortium of pro-bono environmental lawyers.

6 The concept of social metabolism signifies the material and energy throughput flowing in an economy in the form of inputs or waste of the economic activities (see Fischer-Kowalski and Haberl, 2007, 2012:).

7 The data is compiled from various sources including Reinart, 2003; Newspaper Archives; Özkaynak and Rodríguez-Labajos, 2012; Uncu, 2012; yesilartvindernegi. org, 2018; Karahan, 2016.

8 These statistics are released by the Ministry of Energy and Natural Resources.

9 It entailed a mechanism of *urgent expropriation* that expedited land-grabs (for a detailed analysis see Erensü, 2016).

10 Turkey is not the only example that the symbiotic relationship between political authority and capital has unfolded through environmental regulation and mining as well (see for the Cambodia example Sneddon, 2007).

11 It should be noted that the city of Artvin is a peripheral urban space or *taşra* (provincial city) in which socio-ecological rural-urban linkages are rather significant unlike the more globally integrated cities.

Bibliography

Adaman, F., Akbulut, B., and Arsel, M. (2016). Türkiye'de Kalkınmacılığı Yeniden Okumak: HES'ler ve Dönüşen Devlet-Toplum-Doğa İlişkileri. In C. Aksu, S. Erensü, and E. Evren, eds, *Sudan Sebepler: Türkiye'de Neoliberal Su-Enerji Politikaları ve Direnişler*, 1st ed. Istanbul: İletişim Yayınları, pp. 291–312.

Adaman, F., Arsel, M., and Akbulut, B. (2018). Neoliberal Developmentalism, Authoritarian Populism, and Extractivism in the Countryside: The Soma Mining Disaster in Turkey. *The Journal of Peasant Studies*. doi: https://doi.org/10.1080/03066150.2018.15 15737.

Arsel, M. (2012). Opposition to Gold Mining at Bergama, Turkey. *EJOLT Factsheet*, 1st ed. [online] Available at: http://hdl.handle.net/1765/38558.

Avcı, D., Adaman, F., and Özkaynak, B. (2010). Valuation Languages in Environmental Conflicts: How Stakeholders Oppose or Support Gold Mining at Mount Ida, Turkey. *Ecological Economics*, 70(2), pp. 228–238.

Aydın, Z. (2010). Neo-Liberal Transformation of Turkish Agriculture. *Journal of Agrarian Change*, 10(2), pp. 149–187.

Bridge, G. (2004). Mapping the Bonanza: Geographies of Mining Investment in an Era of Neoliberal Reform. *The Professional Geographer*, 56(3), pp. 406–421.

Buğra, A. (1994). *State and Business in Modern Turkey: A Comparative Study*. Albany, NY: State University of New York Press.

Buğra, A. and Savaşkan, O. (2012). Politics and Class: The Turkish Business Environment in the Neoliberal Age. *New Perspectives on Turkey*, 46, pp. 27–63.

Çoban, A. (2004). Community-Based Ecological Resistance: The Bergama Movement in Turkey. *Environmental Politics*, 13(2), pp. 438–460.

Doğru, B. (2017). Cerattepe, Bir Kent Yaşamının Bir Bütün Olarak Tehdit Edildiği İlk ve Tek Vaka. *EkoIQ*, 73, pp. 68–69.

doka.org.tr. (2018). Doğu Karadeniz Kalkınma Ajansı (Eastern Black Sea Development Agency) – *Artvin Provincial Indicators*. [online] Available at: www.doka.org.tr/TR/Bolgemiz/Artvin.

EPA. (2009). *Summary of Key Findings. EPA Toxics Release Inventory, US National Analysis*, 1st ed. Washington, DC: Environmental Protection Agency. [online] Available at:

www.epa.gov/sites/production/files/2018-12/documents/2008_national_analysis_key_findings.pdf.

Erensü, S. (2014a). Murgul'un fendi, Cengiz'i yendi, *Evrensel*. [online] Available at: www.evrensel.net/haber/93320/murgulun-fendi-cengizi-yendi.

———. (2014b). Artvin maden mücadelesini durmadan konuşmamız için 10 neden. . . '. *Evrensel*. [online] Available at: www.evrensel.net/haber/98116/artvin-maden-mucadelesini-durmadan-konusmamiz-icin-10-neden.

———. (2016). Fragile Energy: Power, Nature, and the Politics of Infrastructure in the 'New Turkey'. Ph.D. University of Minnesota.

———. (2018). Powering Neoliberalization: Energy and Politics in the Making of a New Turkey. *Energy Research & Social Science*, 41, pp. 148–157.

Fischer-Kowalski, M. and Haberl, H. (eds). (2007). *Socioecological Transitions and Global Change. Trajectories of Social Metabolism and Land Use*, 1st ed. Cheltenham: Edward Elgar.

Guha, R. and Martinez-Alier, J. (1997). *Varieties of Environmentalism: Essays North and South*. London: Earthscan Publications.

Gümüş, C., Toksoy, D., Eminağaoğlu, Ö., Kurdoğlu, O., Özalp, M., and Turgut, B. (2013). *Artvin İli Cerattepe ve Genya Dağı Ormanlarında Planlanan Madencilik Faaliyetlerinin Doğal Kaynaklar Üzerine Etkileri Hakkında Rapor*. Ankara: Orman Mühendisleri Odası.

Hablemitoğlu, N. (2003). *Alman Vakıfları: Bergama Dosyası*. Istanbul: Toplumsal Dönüşüm Yayınları.

Harvey, D. (2003). *The New Imperialism*. Oxford: Oxford University Press.

Hattam, J. (2015). New solidarity in struggle to protect Turkey's "life spaces". *Mongabay News*. [online] Available at: https://news.mongabay.com/2015/06/new-solidarity-in-struggle-to-protect-turkeys-life-spaces.

Hurley, P. T. and Arı, Y. (2011). Mining (dis) Amenity: The Political Ecology of Mining Opposition in the Kaz (Ida) Mountain Region of Western Turkey. *Development and Change*, 42(6), pp. 1393–1415.

İslamoğlu, H. (2017). The Politics of Agricultural Production in Turkey. In F. Adaman, B. Akbulut, and M. Arsel, eds, *Neoliberal Turkey and Its Discontents: Economic Policy and the Environment Under Erdoğan*, 1st ed. London: I. B. Tauris, pp. 75–102.

İZKA (İzmir Kalkınma Ajansı – Izmir Development Agency). (2013). *Bergama İlçe Raporu*, 1st ed. Izmir. [online] Available at: www.izmiriplanliyorum.org/static/upload/file/2014-2023_ilce_ozet_raporu_-_bergama.pdf.

Kalın, B. (2016). Interview with Avukat Bedrettin Kalın for *Express*, 142, pp. 54–56.

Karadag, R. (2010). Neoliberal Restructuring in Turkey: From State to Oligarchic Capitalism. *Max Planck Institute for the Study of Societies Discussion Paper*, 10(7).

Karadeniz, S. and Kalın, B. (2016). *Cerattepe Maden Dosyası*, 1st ed. Ankara: Artvin Kültür ve Yardımlaşma Derneği; Artvin: Yeşil Artvin Derneği. [online] Available at: http://yesilartvindernegi.org/cerattepe-maden-dosyasi/.

Karahan, N. N. (2016). Interview with Nur Neşe Karahan for *Express*, April, pp. 52–56.

———. (2017). Interview with Nur Neşe Karahan for *Bianet*, August 16.

Keyder, Ç. (1981). *The Definition of a Peripheral Economy: Turkey 1923–1929*. New York: Cambridge University Press.

Keyder, Ç. and Yenal, Z. (2011). Agrarian Change Under Globalization: Markets and Insecurity in Turkish Agriculture. *Journal of Agrarian Change*, 11(1), pp. 60–86.

Konak, N. (2008). Ecological Modernization and Eco-Marxist Perspectives: Globalization and Gold Mining Development in Turkey. *Capitalism Nature Socialism*, 19(4), pp. 107–130.

Krausmann, F., Gingrich, S., Eisenmenger, N., Erb, K. H., Haberl, H., and Fischer-Kowalski, M. (2009). Growth in Global Materials Use: GDP and Population During the 20th Century. *Ecological Economics*, 68(10), pp. 2696–2705.

Latorre, S., Farrell, K. N., and Martínez-Alier, J. (2015). The Commodification of Nature and Socio-Environmental Resistance in Ecuador: An Inventory of Accumulation by Dispossession Cases, 1980–201. *Ecological Economics*, 116, pp. 58–69.

Martinez-Alier, J. (2001). Mining Conflicts, Environmental Justice, and Valuation. *Journal of Hazardous Materials*, 86(1–3), pp. 153–170.

———. (2002). *The Environmentalism of the Poor: A Study in Ecological Conflicts and Valuation*. Cheltenham; Northampton: Edward Elgar.

———. (2009). Social Metabolism, Ecological Distribution Conflicts, and Languages of Valuation. *Capitalism Nature Socialism*, 20(1), pp. 58–87.

———. (2012). Environmental Justice and Economic Degrowth: An Alliance Between Two Movements. *Capitalism Nature Socialism*, 23(1), pp. 51–73.

Martinez-Alier, J., Temper, L., Del Bene, D., and Scheidel, A. (2016). Is There a Global Environmental Justice Movement?. *The Journal of Peasant Studies*, 43(3), pp. 731–755.

Marx, K. (1973). *Grundrisse*. New York: Vintage Books.

Moore, J. W. (2000). Sugar and the Expansion of the Early Modern World-Economy: Commodity Frontiers, Ecological Transformation, and Industrialization. *Review*, 23(3), pp. 409–433.

Muradian, R., Walter, M., and Martinez-Alier, J. (2012). Hegemonic Transitions and Global Shifts in Social Metabolism: Implications for Resource-Rich Countries. *Global Environmental Change*, 22(3), pp. 559–567.

Nixon, Rob. (2011). *Slow Violence and the Environmentalism of the Poor*. Cambridge, MA; London: Harvard University Press.

Olson, D. and Dinerstein, E. (2009). The Global 200: Priority Ecoregions for Global Conservation. *Annals of the Missouri Botanical Garden*, 89, pp. 199–224

Öncü, A. (2014). Turkish Capitalist Modernity and the Gezi Revolt. *Journal of Historical Sociology*, 27(2), pp. 151–176.

Orhan, G. (2006). The Politics of Risk Perception in Turkey: Discourse Coalitions in the Case of the Bergama Gold Mine Dispute. *Policy & Politics*, 34(4), pp. 691–710.

Özen, H. (2009). Located Locally, Disseminated Nationally: The Bergama Movement. *Environmental Politics*, 18(3), pp. 408–423.

Özen, H. and Özen, Ş. (2011). Interactions in and Between Strategic Action Fields: A Comparative Analysis of Two Environmental Conflicts in Gold-Mining Fields in Turkey. *Organization & Environment*, 24(4), pp. 343–363.

Özen, Ş. and Özen, H. (2009). Peasants against MNCs and the State: The Role of the Bergama Struggle in the Institutional Construction of the Gold-Mining Field in Turkey. *Organization*, 16(4), pp. 547–573.

Özgökçeler, S. and Sevgi, H. (2016). Cerattepe: As an Explanandum of the Common Faith. *The Journal of International Social Research*, 9(45), pp. 501–507.

Özkaynak, B. and Rodríguez-Labajos, B. (2012). Mining Conflicts Around the World: Common Grounds from an Environmental Justice Perspective. *Ejolt Report*, 7. Available at: www.ejolt.org/2012/11/mining-conflicts-around-the-world-common-grounds-from-an-environmental-justice-perspective/.

Perreault, T. (2013). Dispossession by Accumulation? Mining, Water and Nature of Enclosure on the Bolivian Altiplano. *Antipode*, 45(5), pp. 1050–1069.

Reinart, Ü. B. (2003). *Biz Toprağı Bilirik: Bergama Köylüleri Anlatıyor*. Istanbul: Metis Yayınları.

Rodríguez-Labajos, B. and Özkaynak, B. (2017). Environmental Justice Through the Lens of Mining Conflicts. *Geoforum*, 84, pp. 245–250.

Schmidt, A. (1965). *The Concept of Nature in Marx*. Bristol: Western Printing Services.

Şekercioğlu, Ç. H. (2011). *Turkey's Conservation Crisis: Global Biodiversity Hotspots Under Threat*. [online] National Geographic. Available at: https://blog.nationalgeographic. org/2011/12/31/turkeys-conservation-crisis-global-biodiversity-hotspots-under-threat/.

Şekercioğlu, Ç. H., Anderson, S., Akçay, E., and Bilgin, R. (2011a). Turkey's Rich Natural Heritage Under Assault. *Science*, 334(6063), pp. 1637–1639.

Şekercioğlu, Ç. H., Anderson, S., Akçay, E., Bilgin, R., Can, Ö. E., Semiz, G., Tavşanoğlu, Ç., Yokeş, M. B., Soyumert, A., İpekdal, K., Sağlam, İ. K., Yücel, M., and Dalfes, N. (2011b). Turkey's Globally Important Biodiversity in Crisis. *Biological Conservation*, 144(12), pp. 2752–2769.

Sneddon, C. (2007). Nature's Materiality and the Circuitous Paths of Accumulation: Dispossession of Freshwater Fisheries in Cambodia. *Antipode*, 39(1), pp. 167–193.

Sneddon, C. and Fox, C. (2012). Inland Capture Fisheries and Large River Systems: A Political Economy of Mekong Fisheries. *Journal of Agrarian Change*, 12(2–3), pp. 279–299.

Tamzok, N. (2005). Türkiye Madencilik Sektöründe Yapısal Dönüşüm ve Sonuçlar. *Turkey 19. International Mining Congress, Izmir, Turkey, 9–11 June*. Ankara, TMMOB, pp. 5–20.

Tsing, A. (2003). Natural Resources and Capitalist Frontier. *Economic and Political Weekly*, 38(48), pp. 5100–5106.

tuik.gov.tr. (2018). *Address Based Population Registration System*. [online] Available at: https://biruni.tuik.gov.tr/medas/?kn=95&locale=tr.

Turhan, E. (2017). Right Here, Right Now: A Call for Engaged Scholarship on Climate Justice in Turkey. *New Perspectives on Turkey*, 56, pp. 152–158.

Uncu, B. A. (2012). *Within Borders, Beyond Borders: The Bergama Movement at the Junction of Local, National and Transnational Practices*. Ph.D. London School of Economics.

The Union of Chambers of Turkish Engineers and Architects (TMMOB). (2014). *TMMOB Cerattepe Raporu*, 1st ed. Ankara: TMMOB. [online] Available at: www. tmmob.org.tr/sites/default/files/cerattepe.pdf.

Urkidi, L. and Walter, M. (2011). Dimensions of Environmental Justice in Anti-Gold Mining Movements in Latin America. *Geoforum*, 42(6), pp. 683–695.

Watts, M. (2015). Securing Oil. Frontiers, Risk and Spaces of Accumulated Insecurity. In H. Appel, A. Mason, and M. Watts, eds, *Subterranean Estates: Life Worlds of Oil and Gas*, 1st ed. Ithaca: Cornell University Press, pp. 210–236.

Yaşın, Z. (2017). The Adventure of Capital with Nature: From the Metabolic Rift to the Value Theory of Nature. *The Journal of Peasant Studies*, 44(2), pp. 377–401.

yesilartvindernegi.org. (2018). *Yeşil Artvin Derneği – Mücadele Tarihimizden*. [online] Available at: http://yesilartvindernegi.org/mucadele-tarihimizden/.

9 Coal, ash, and other tales

The making and remaking of the anti-coal movement in Aliağa, Turkey

Ethemcan Turhan, Begüm Özkaynak, and Cem İskender Aydın

Situated 50 kilometers north of Turkey's third-largest city, Izmir, Aliağa is home to shipbreaking and smelting facilities, oil refineries and massive coal-fired power plants. Aliağa Bay – located on the Aegean coast, with abundant scenic landscapes, pristine waters, and archaeologically important sites – was initially designated as a heavy industrial development zone by the 1961 Constitution. This was followed by the establishment of state-owned heavy industries, particularly during the 1980s; namely, PETKİM (petrochemicals) and TÜPRAŞ (oil refinery), despite the potential to develop tourism in the region. Small and medium-scale industries, such as shipbreaking, iron-steel smelting, and cement manufacturing flourished around these two large state-owned facilities, complementing them and serving the domestic and international strategic interests of Turkish governments and industrial groups. Industrial clustering around iron, steel, and cement was later supplemented with fossil fuel–based energy production facilities. Accompanying the years of state-led industrialization, a strong working class grew alongside the facilities in the region. The lack of cumulative impact studies coupled with a diverse set of state-led polluting investments was influential in turning Aliağa and its environs into an "ecological sacrifice zone" (Lerner, 2010). Today, approximately 36 percent of Turkey's crude oil is processed in Aliağa, and ambient levels of volatile organic compounds (VOCs) are four to 20 times higher than suburban locations in the Izmir metropolitan area (Çetin et al., 2003). Cancer risk is high in the region due to these pollutants, at four times the levels considered acceptable by the U.S. Environmental Protection Agency (Civan et al., 2015).

The Aliağa region has a tumultuous history of social struggles stretching over the past 40 years, with the rise and demise of working-class action against large-scale privatizations, as well as a fierce environmental movement propelled by the local community in tandem with local authorities and national/international networks. One climactic point was the 50,000-strong human chain in Aliağa on May 6, 1990, to protest the planned imported coal-fired power plant. On May 15, 2016, some 26 years later after this fateful campaign, Aliağa became home to a second mass mobilization against coal-fired power plants and coal ash dumpsites. However, this time the framing, repertoire of contention, political context, and the alliances of the movement were considerably different from the first mobilization, with climate change being a major part of the contemporary anti-coal narrative. Government plans to expand liquefied natural gas (LNG) terminals and allow additional coal-fired power capacity and associated ash residue dumpsites on the watershed of nearby villages continue to cause significant dissent among the residents, particularly in Yeni Foça, a district that overlooks Aliağa Bay.

In this chapter, we take a critical look at the historical transformation of grassroots mobilization and political engagement in Aliağa in the period between these two historical moments (1990 and 2016) by using archival material from two national newspapers with wide circulation, secondary literature, and in-depth interviews with some of the key actors. Aliağa appears to be a curious case for neglect in the scholarly literature on environmental activism in Turkey, a history of victories and defeats only partially told. This is particularly relevant and important since the powerful coalition that had emerged in the 1990s (formed by locals, the Green Party [*Yeşiller Partisi*], the main social democratic opposition party in parliament, the Union of Chambers of Turkish Engineers and Architects as well as labor unions) fought and won a major victory giving way to the cancellation of the government's plans and the birth of a combatant environmental movement in the region. Although it was one of the first nationally debated environmental justice successes of this scale in Turkey (Şahin, 2010), anti-coal movement in Aliağa still remains somewhat under-investigated in the country's history of environmental movements. Thus, providing a micro-historical account would not only give the Aliağa anti-coal movement the due credit it deserves, but also help us illustrate the changing nature and shifting contours of environmental mobilizations in Turkey at large in a time of re-escalating authoritarianism. Since "there is not a right or wrong environmentalism, but narratives and practices of environmentalism which are historically produced" (Armiero and Sedrez, 2014: 11), our effort here also helps to reveal some hidden narratives and practices which are equally relevant for contemporary environmental movement in Turkey. To this end, we describe how the hegemonic state – in a counter-movement – reacted to the legal developments and the activism in Aliağa by changing the rules of the game; amending institutional and legal frameworks for investment decisions as needed, thereby speeding up and deepening neoliberal reforms. The tale of the anti-coal struggle in Aliağa presented in this chapter is important for environmental struggles in general, as it offers interesting insights into the ways environmental movements and their counter-hegemonic powers confront, clash, and negotiate with the state just to die out and eventually be reborn.

In terms of research methodology, we coded and analyzed a total of 859 newspaper clippings from two major national newspapers (*Milliyet* and *Cumhuriyet*) and categorized the data into three periods: 1980–1994 (431 clippings), 1995–2004 (128 clippings), and 2005–2015 (300 clippings). We also visited the site several times and conducted multiple interviews with anti-coal movement members. Collating the empirical data, secondary literature, and interviews provided us a rich source of material from which we drew results. The three time periods are strikingly different phases at the national and regional scales, which are all highly relevant for the anti-coal movement in Aliağa. The first period, from late 1980s to the 1994 economic crisis, corresponds to the first stage of neoliberal restructuring in Turkey. The first energy investment in the Aliağa region – and the ensuing mass mobilization/resistance – also took place during this period. The second period, between 1995 and 2005, is significant due to its coalition governments and the continued albeit slower effects of neoliberal reforms, resulting in political and economic instability and the subsequent economic crises of 1999 and 2001 in Turkey. These political and economic failures were critical in that they were followed by the rise of Recep Tayyip Erdoğan's Justice and Development Party (AKP; *Adalet ve Kalkınma Partisi*) gaining power in the 2002 elections, which has been the dominant political force in Turkey since then. This latter period is also noteworthy for the new environmental struggles that emerged in Turkey, initially related to mining and hydro-power but now increasingly related to energy metabolism expansion more broadly, crucial to understanding why and how the anti-coal movement in Aliağa was sidelined (Adaman et al., 2017). Finally, the post-2005 period was characterized by a new wave of neoliberal economic reforms led by the powerful AKP regime and mass privatization. This is also the period in which new coal-fired power plant investments loomed large over Aliağa, triggering combatant anti-coal reaction anew. In this context, *Yeni Foça Forum* emerged as a new actor in the region as an offshoot of the 2013 Gezi Park Protests (Özkaynak et al., 2015) and the 2016 Break Free from Fossil Fuels action.[1]

In what follows, we first introduce the early phase of the anti-coal movement in Aliağa in two sections that cover the periods from 1980 to 1990 and from 1990 to 1995, respectively. Then, the third section documents the phase between victory and defeat: the 1995–2005 period. An account of the early years of AKP rule and the new dynamics of local struggle in the region between 2005 and 2016 follows in the fourth section. The chapter then concludes by offering some ideas for a synthesis of the continuities and ruptures of the environmental struggle in Aliağa.

When foreign coal comes to town (1980–1990)

> *Turkey at one stroke left behind the Third World evolutionary phase and entered a new one full of the promises and challenges of modern industrial society.*
> —Former Prime Minister Turgut Özal (1987)

Coming out of an iron-fisted coup d'état at the beginning of the 1980s, which not only crushed the political left but also enabled and secured the rapid neoliberal transformation of the country's import-substituting economy, Turkey witnessed radical market-oriented reforms in a largely authoritarian setting throughout the decade (Öniş, 2004; Yalman, 2009; Tonak and Akçay, 2019). Together with the military cadre that led the September 12, 1980 coup d'état, which maintained its control over society for at least two more decades, former Prime Minister (later

President) Turgut Özal was without a doubt the key figure at a time when the country was opening its assets to foreign investors. Özal, himself coming from a technocratic career in State Planning Organization (DPT, *Devlet Planlama Teşkilatı*) with a degree in electrical engineering, made clear that his government would be prioritizing major energy infrastructure investments after he took office in November 1983. About a year into Özal's rule, the first clues of what was in the making were revealed. As of 1984, less than a quarter of the country's primary energy needs for electricity production was import-dependent (TMMOB, 2016). This was also the time when rumors first appeared in mainstream newspapers of the government's plans for three thermal power plants that will run on imported coal. On November 13, 1984, *Milliyet* columnist Mümtaz Soysal (a professor of constitutional law and later Minister of Foreign Affairs) commented:

> No one says energy shouldn't be produced. Neither does one say no to thermal power plants. We regard the pollution in the vicinity of Çatalağzı, Aliağa and even Silifke as the presently inevitable cost of producing an industrial Turkey. All that is said is this, indeed: Do we have to put up a thermal power plant in a place with incredible beauty that needs to be protected as a national park for the future generations?
>
> (Milliyet, November 13, 1984)

This initial outcry was about the Gökova thermal power plant (completed in the 10 years between 1983 and 1993) running on domestic lignite (630MW), a political move that sparked a sizeable popular opposition. However, there were much larger plans for energy infrastructure to come.

"*Turkey will be 40 percent better lit*" announced a map published in *Milliyet* (July 7, 1985) showing the approximate locations of three imported coal-fired power plants. Shortly thereafter, the Özal government contracted American company Bechtel to conduct initial feasibility studies for a 600MW plant in Aliağa. Consequently, the remarks from a high-level bureaucrat from State Planning Organization showed what would dominate the newly forming environmental movement's agenda for the coming years: "*It is not possible for the coastlines to be polluted by these plants. If we keep saying these kinds of things, it will damage the investments. The Italian-French consortium wants to invest in the already polluted region of Aliağa*" (Milliyet, August 20, 1986). Now evident, Özal's grand plan was to construct 21 imported coal-fired power plants and 37 domestic lignite-powered plants between 1993 and 2010. In an op-ed to *Washington Quaterly* in 1987, he made his intentions clear: "*Here again, we have elaborated a new mode of investment which we call 'build-operate-transfer (BOT), and if you like, hand over'. This approach is accepted by investors and their countries. As a first step, some larger thermal power plants are on the way to realization*" (Özal, 1987: 164). Bandwagoning the global narrative of looming energy scarcity from the early 1980s, the Turkish government's simultaneous move to open up the country to global market capitalism and encourage foreign direct investment in the energy sector came to the attention of local authorities in Aliağa and the neighboring town of Yeni Foça at the beginning of 1987. Designated as a heavy industrial zone following the constitutional change in 1961, Aliağa was already host to a number of polluting facilities including chemical and petrochemical industries, refinery facilities, smelters, and a shipbreaking yard starting from the 1970s and building up further in the first half of the 1980s. After a visit to Ankara, former Aliağa mayor İrfan Onaran reported that "*They are completely sacrificing our region for the sake of a few industries. It is often forgotten that our region is also a tourism zone. There is a French holiday resort right next to the proposed power plant site. We certainly do not want this power plant*" (*Cumhuriyet*, January 17, 1987).

In a time when the labor movement was slowly re-organizing itself after the coup d'état and the new social movements were coming of age, Aliağa swiftly became a site of political contestation over the energy sector. Over the next year, it was slowly revealed that the government had been secretly planning to expand industrial activity, through a coal power plant project constructed and maintained by a Japanese company. This revelation almost immediately led to an unrest in the community. In the summer of 1988, Kemal Anadol, an opposition MP from Izmir and a labor rights lawyer, was responding to local concerns over the designation of a coastal village, Gencelli, during a regular town hall. In his book aptly titled *No to Thermal Power Plants*, he gave a surprising account of his meeting with a local resident:

> *A citizen came by my side and said "Kemal Bey, please take a look at this map. There is a coal-fired power plant here. This is the important point. The Japanese will be constructing a large plant here. Electricity produced by this plant will be used by the existing and planned privately-owned steel factories. There are rumors that some politicians and their entourage are involved in this."*
>
> (Anadol, 1991: 19)

Anadol took this issue personally and dug into it. The first instances of popular grievances appeared in the media in late 1988, when the then-Minister of Public Works Sefa Giray stripped away the authority of Aliağa and Foça municipalities resisting the smelter facility in the vicinity of Gencelli. Over the next few months in the run-up to March 26, 1989, local elections, the grievances about the complicity of local authorities from Özal's governing party further accelerated and eventually gave way to a landslide victory of the Social Democratic Populist Party (SHP; *Sosyaldemokrat Halkçı Parti*) across 19 municipalities in the region. After the elections, the incoming Minister of Public Works and Housing Cengiz Altınkaya was quoted as saying: "*This is not the only industry being built in the bay. There are 4–5 iron-steel facilities, 28 shipbreaking yards and an oil refinery there. It is already a lost region, what would matter if another factory was built?*" (*Yeni Asır*, June 9, 1989).

The case of the coal-fired power plant got even bigger attention when newspapers started quoting anonymous state officials that the project would be implemented by the Japanese energy utility, EPDC, and coordinated not by the Turkish Electricity Authority (TEK; *Türkiye Elektrik Kurumu*) (formally in charge of energy investments) but by the DPT's Department of Foreign Direct Investments. Recognizing the gathering political storm, 12 mayors of neighboring municipalities all of whom were from the Social Democratic Populist Party gathered in Gencelli on August 25, 1989, where the residents staged an impromptu road-block with banners reading: "*We don't want thermal power plant, we want nature to thrive.*" While smaller instances of anti-coal fired power plant activism, mainly led by women, continued, a new phase of the environmental movement kicked off on September 12, 1989, when the State Minister in charge of economic affairs, Güneş Taner, publicly, and finally, confirmed that a thermal power plant running on imported coal would be built by Japanese investors. This declaration became the last nail in the coffin, which then led to the rapid consolidation of Aliağa's anti-coal movement as an unlikely alliance of diverse actors across the country, which was still recovering from the authoritarian regime of the 1980s. Actions became "joyful repertoires of contention" (Della Porta, 2013), with complementary "critical mass"–style direct actions with bicycles and establishing guard posts (November 18, 1989), support visits by famous musicians (December 10, 1989), and even pantomime acts for the villagers (December 11, 1989). The Izmir branch of the Green Party and SOS Akdeniz (a regional NGO) spearheaded the movement by using their media visibility and establishing contacts with international partners from Greece and Germany. They also established a series of popular mass actions that dominated public agenda in Izmir. Ecologist and activist Savaş Emek, a highly influential figure in the Aliağa anti-coal movement and the then-provincial representative of the Green Party, referred to the consolidation of the anti-coal movement in this period as follows: "*The actions [in Aliağa] were all planned step by step. Not everything was done all together. For me, this is the advantage of being a veteran of the socialist tradition, knowing to work in a planned fashion, escalating the fight step by step. We used this to our advantage*" (Şahin and Mert, 2006). Escalating the anti-coal struggle, nonetheless, required a significant coordination not only among actors but also among demands. Thus, this period is also significant for mobilizing "polluter pays" (*Milliyet*, December 17, 1989) arguments conforming with those of global environmentalist movements, in tandem with more radical environmental justice claims such as "*Aliağa is treason against humanity*" (*Milliyet*, May 4, 1990).

Consolidating an anti-coal movement (1990–1995)

> One of the Turkish state's fears that surfaced during Aliağa [struggle] was this: If these people rise up against coal-fired power plants, this will make a precedent. Then where would we build power plants?
>
> —Ecologist and activist, Savaş Emek (2015: 135)

The signal flare of what would later be referred to as "*the gospel of getting over with being a silenced community*" by former mayor of Izmir, Yüksel Çakmur (*Milliyet*, May 7, 1990), in Aliağa was effectively lit in the fall of 1989. Partnering up with a local neighborhood association in Foça, the Turkish Chamber of Mechanical Engineers (MMO; *Türkiye Makine Mühendisleri Odası*) issued a press release in September 1989 emphasizing that Turkey already had an installed capacity of 75 billion kWh of electricity whereas it could only consume 50 billion kWh of it. The logic of build-operate-transfer (BOT) model, the new economic leverage of the Özal government to attract foreign direct investment, was not holding much ground among the local population. Thus, the state institutions decided to play more aggressively. The head of DPT's Department of Foreign Direct Investments, İbrahim Çakır, first suggested that "*there will be no more bureaucratic hurdles to stop Aliağa thermal power plant*" (*Cumhuriyet*, October 30, 1989) and then upped the ante in the face of mounting opposition, saying "*the Aliağa thermal plant, which will be built by 70 percent Japanese capital will not cause any environmental damage*" (*Cumhuriyet*, November 24, 1989). While the popular opposition was growing, the Council of Ministers issued a governmental decree on October 18, 1989, officially announcing the establishment of a joint venture company (70 percent Japanese, 30 percent Turkish capital) for the construction and operation of the proposed power plant. The key legal trick here was the use of the free trade zone law, which in essence was meant to facilitate land allocation for export-oriented purposes. Yet somewhat contradictorily, the Aliağa-Gencelli power plant would become the country's first plant running on imported coal, burning coal arriving from places as far as Australia, South Africa, and Colombia as well as being mainly owned by foreign investors. Unsurprisingly perhaps, the emerging opposition was not easy to convince about the benefit of the plan and thus came the storm of court cases led by lawyers from Izmir Bar Association (*İzmir Barosu*) and Kemal Anadol himself to the Council of State (*Danıştay*). Anadol would later refer to this legal move as the "*never-ending fight*," the first instance of organized citizen reaction in the aftermath of the bloody 1980 coup (Anadol, 1991: 35).

As the legal fight was gaining steam, the mayors of 12 municipalities in the region organized under the umbrella of the Bakırçay Municipalities Union (*Bakırçay Belediyeler Birliği*) and started collaborating with activist groups. The first mass act of opposition from this group came in early November 1989 in the shape of a referendum on the thermal power plant. The Aliağa mayor, Hakkı Ülkü, described this move as "*the first urban citizen referendum in republican history*" (*Cumhuriyet*, November 16, 1989). The results of 7,717 votes cast were self-explanatory with 94 percent "No!" response in what could be referred to as the first act of direct democracy on an environmental matter in the country (Figure 9.1). Consequently, the outcome did not go uncontested since the Ministry of Interior immediately started a formal investigation into the Bakırçay Municipalities Union for extra-legal use of authority in organizing a popular referendum without the central government's consent. Feeling the growing dissent, local organizers and SOS Akdeniz also reached out to Greek social democrats, thereby leading three Panhellenic Socialist Movement (PASOK) deputies to pose a parliamentary question to the Greek Minister of Interior on the potential impacts of coal-fired power plants on the island of Lesvos right across the Aliağa Bay (*Cumhuriyet*, December 10, 1989). By the end of the year, the undersecretary

Figure 9.1 Popular referendum on coal-fired power plants organized by Foça Municipality on November 15, 1989

Source: Photo courtesy of Ümit Otan

for the environment, Zeynep Arat, commented that "*there has never been an environmental assessment [in Aliağa]. We need to be smart and ask for a package deal. Otherwise, they will first sell us the power plant and then the [waste] treatment facilities*" (*Cumhuriyet*, December 4, 1989).

In a parliamentary session on February 28, 1990, Kemal Anadol took the floor to have a heated debate with the Minister of Energy, Fahrettin Kurt, who appeared to be the fiercest defendant of the proposed coal-fired power plant in Aliağa. Hidden between the lines was that the coal ash produced as a result of burning imported coal was to be transferred to an ash dump site 3 km away from the plant in closed vessels. The government argued that this ash would be stored there for 26 years before the zone was to be rehabilitated and transformed into a recreational zone (or even an agricultural zone in some accounts) (Anadol, 1991: 67–76). Moreover, the Minister was openly admitting that Aliağa was selected as "*it was designated as an industrial zone and already hosted 27 industries*," clearly neglecting the possible cumulative impact of these industries (Ibid.: 72). However, it was not only the Turkish government that was deeply concerned by the growing distaste with the project but also the Japanese company EPDC so as to even prompted an official visit by the Japanese Prime Minister (*Milliyet*, April 30, 1990).

All these eventually culminated in the emblematic direct action on May 6, 1990. Benefiting from the support of Izmir Metropolitan Municipality and Bakırçay Municipalities Union, the organizing committee led a 50,000-people strong human chain action. With participants from cities as far as Adana, Samsun, and Trabzon, the human chain (also dubbed "Love Chain") action sparked much larger attention despite the fact that the state-run TV channel TRT broadcasted a documentary film greenwashing Japanese coal-fired power plants the night before the action (*Cumhuriyet*, May 7, 1990). Radical demands on banners such as "*We are not voters anymore, we are citizens*," "*Bleachers to the fields, democracy to the streets*," and "*Coal-fired power plants are the enemies of humanity*" were important signposts showing the narrative reach of the movement (Ibid.). The good news finally arrived on May 8, 1990, two days after the mass action, when the Council of State announced a stay of execution decision for the proposed power plant. While the victory celebrations were on, nonetheless, the state officials were quick to declare that "*the investment plans were not annulled*" (*Milliyet*, May 8, 1990). With an emergency decree on May 10, 1990, after an extraordinary meeting between the Japanese company and President Özal, the government opened another legal channel for the investment by expanding the borders of the Aliağa free trade zone. Confronting Kemal Anadol in parliament the following day, the Minister of Public Works, Cengiz Altınkaya, commented that "*the street protests will not change our determination. If, as the government, we allow things to be handled by the streets then we would have to give up on all power plants in Turkey*" (*Milliyet*, May 11, 1990).

Emboldened by the initial stay of execution decision from the higher court, the legal fight in Aliağa accelerated along with the reciprocal war of words between the anti-coal movement and the state. Furthermore, desertion of the Minister of Tourism İlhan Aküzüm to anti-coal ranks gave further impetus to this legal-institutional component of the movement. After a legal ping pong that made the Japanese counterparts anxious, a company spokesperson even shifted to black-mailing: "*Aliağa needs to be finished as initially planned. After all, this is your problem. We would wait a bit more but not much. It is hard to invest in Turkey. Abandoning Aliağa* [power plant project] *would have unpleasant side effects. Maybe there was not much opposition, they were not so numerous but they have been effective. The fate of this project will be the benchmark for future Japanese investments*" (*Milliyet*, January 21, 1992). Yet, the times they were a-changin'. After 10 stay of execution decisions over two years, the Council of State finally annulled the second decree of the government on April 28, 1992, on the grounds of "*ecological equilibrium*." This ruling prompted different legal interpretations from lawyers and environmental activists attributing a moral higher ground to the court by suggesting that its decision was to "*put ecology before the national interest*," although not everyone was in agreement on this (*Cumhuriyet*, May 5, 1992). Eventually, it was Prime Minister Süleyman Demirel who knocked down the project with a flamboyant press statement on his way back from the Rio Earth Summit in 1992, where three major international environmental agreements (UNFCCC, UNCCD, and UNCBD) were launched. Riding the global wave of environmental optimism,

Demirel commented that the Aliağa coal-fired power plant project will not continue and even went as far as to call the ongoing coal-fired power plant construction in Gökova *"murder"* (*Cumhuriyet*, June 16, 1992). Kemal Anadol would later refer to this victory as the success of *"a civilian movement which will end the ill-fate of Turkey and put an end"* to the authoritarian regime marked by the March 12, 1971, and September 12, 1980, coup d'états (*Milliyet*, June 21, 1992). Disappointed by the cancellation, the Japanese EPDC started to seek compensation for its *"10 million USD loss"* (*Cumhuriyet*, June 22, 1992).

Between victory and defeat (1995–2005)

> *The case of the power plant in İzmir Aliağa is a pity. That was a green power plant but didn't suit some people's book. In fact, Turkey will be facing energy scarcity in 1995–96.*
> —Former president Turgut Özal (*Milliyet*, August 8, 1992)

The legal turn of the Aliağa anti-coal movement proved to be a winning card in the 1990s, mainly because it was not the solitary effort of a single organization, rather, it brought together a diverse set of actors that formed a broad supporter base (including municipalities, citizen groups, NGOs, academics, and professional chambers). Consequently, one of the victories of the movement in Aliağa was this merger of different opposition groups over an environmental justice claim. According to one of the lawyers of the movement, this was a somewhat organic process: *"As the movement strengthened, there was more press coverage, which further enhanced participation in the movement, and everyone began to express themselves in that environmental movement"* (Interview on April 24, 2017). Along these lines, even the Aegean Region Chamber of Industry (EBSO; *Ege Bölgesi Sanayi Odası*) – an important regional actor for industry representatives – eventually had to take the claims of the movement seriously and participate in local meetings held a number of times to better comprehend the communities' environmental concerns. It is therefore remarkable that a formal mediation process including 62 different stakeholders in Aliağa was conducted between December 1996 and May 1997 (Müezzinoğlu, 2000). Despite the carefully designed process exploring multiple contested issues (including land use, pollution sources, air-water-soil quality, and new energy projects), the final results, including the use of long-term cumulative strategic EIAs (Environmental Impact Assessment), were *"not reflected in a definite management or implementation program"* (Ibid.: 56).

Needless to say, there were also certain national and global circumstances that supported the anti-coal movement and helped bring about the successful outcome of stopping the coal-fired power plant. Looking at these structural influences together with the local factors is helpful in explaining why a communitarian gathering, followed by a legal victory, happened then and there. Our analysis identified three sets of interrelated forces that facilitated this collective outcome. The first set was the post-1980s political atmosphere in Turkey, where personal and political freedoms were expanding and civic mobilization was (re) gaining force around issues such as gender, human rights, and the environment in the aftermath of the brutal military coup that severely curtailed democratic rights and mechanisms for political participation. Our respondent from the Aliağa anti-coal movement also echoed this sentiment and said, "*The Özal administration's attitude towards this local movement was not harsh. There was a tolerant, liberal environment*" (Interview on April 24, 2017). The second set of concerns included the neoliberal restructuring and governance trends at the global level that were quite new and assumed an increasingly significant role for civil society participation (as also emphasized by the Rio Earth Summit in 1992 on the environmental front), and hence still allowed for some push back with regard to privatization and deregulation. Also of note is the fact that neoliberal ideology in Turkey had not yet become as influential in the broader social field in the early 1990s. At the time, governments were still relatively slow with the privatization program as they did not want to lose full control, labor and trade unions were still around and alive, and the size and depth of the capital market was limited. Finally, the third set of forces was related to the international popularity of the sustainable development and Local Agenda 21 discourse again in the 1990s, which put pressure on national states with regard to both local and international environmental issues.

While Prime Minister Demirel's statement discarding the project at first seemed like great news to Turkish environmentalists, counter-statements from the state officials arrived quickly – and perhaps unsurprisingly. For instance, the General Director for Environmental Impact Assessment under the Ministry of Environment, Murat Sungur, commented that "*cessation of the project was not due to it being a coal-fired power plant but due to the legal hurdles related with the free trade zone*" (*Cumhuriyet*, July 5, 1992). Since this period also coincided with the time that the consolidating environmental movement in Turkey was throwing its weight behind the anti-nuclear struggle in Akkuyu and anti-gold mining in Bergama, the new conservative government was therefore even more adamant to suggest that they will "*launch an intense campaign to deploy a counter pro-thermal narrative in the media to prevent environmental reactions*" (*Cumhuriyet*, November 10, 1996).

Overall, throughout the 1990s the burgeoning environmental movement was keen on bolstering the power of local agency through cooperation and networks, such as the alliance between labor unions and environmentalists, strong relations with international counterparts, and collaborations with academia. As a result, environmental resistance in the Aegean region continued on diverse fronts in the 1990s (industrial pollution and oil spills, shipbreaking, rapid urbanization in coastal areas, overfishing and the protection of seals) with mainly non-violent strategies (court appeals, appeals against the environmental impact assessment, alternative reports, data collection of health impacts and collaboration between scientists and activists, workers' festivals, petition campaigns, marches from Izmir to Aliağa to Bergama, Greek-Turkish environmental meetings, etc.). Human rights, democracy, and rule of law were also among the largely credible and widely used themes in opposition discourses generated in reference to the anti-democratic practices of the governments that served throughout the decade. Indeed, the spectrum of such discourses and tactics reflect a continuum between resistance strategies of the environmental movements in Turkey in the 1990s (Orhan, 2006; Arsel, 2012; Özen and Özen, 2018). However, the rapid urbanization and the domination of middle-class values in this period also undermined the possibilities of an alternative political strategy that contrasted clearly to the mainstream politics ferociously criticized by the environmental movement (Özlüer et al., 2016).

In the 1990s, one particular mobilization in close vicinity to Aliağa gained nationwide public sympathy. It took place in Ovacık, Bergama, against the operations of Eurogold, a multi-national company and subsidiary of Australia's Normandy Mining Ltd. (see Yaşın, in this volume). This was the very first anti-gold mining mobilization following the opening of the Turkish mining to foreign investments in 1985 in line with liberalization policies. While the struggle of the villagers did attract popular attention, the anti-mining discourse's emphasis on Eurogold being a foreign company to some extent overshadowed the local movement's stand against the negative environmental effects of mining operations and, perhaps more importantly, the broader neoliberal capitalist structure that gives way to such operations by both national and foreign companies (Özen and Özen, 2018). Moreover, the unexpected popularity of this movement somewhat eclipsed the success of the Aliağa struggle and diverted attention from energy investment–related controversies to gold-mining operations and cyanide pollution in the region, and to imperialism at large.

One key weakness of the protest movements in this period was their lack of imagination in developing and explicitly articulating an alternative vision in the face of foreign capital investments that promised local and national economic development. Preoccupied with continuous daily shocks and struggles, these movements – while successful in expressing their discontent with the proposed projects and plans – had neither enough time nor energy to situate their discourses in a positive framework. This limited their capability to counter the dominant discourses on development and national interests, which were seen as strongly tied to the economic contributions these projects would generate for the region and the country. In Aliağa, the hegemonic modernist discourse centered on looming energy scarcity from the 1980s onwards; offsetting this discourse and offering alternative energy policies and management practices would have been crucial for the long-term viability of the resistance.

Somewhat counter-intuitively, the tale of Aliağa does not end with the local people's victory and living happily ever after. The fact that the state did not grant an excavation permit for the ancient city of Kyme in the outskirts of Aliağa was an early signal of its reluctance to protect the region's historical and ecological wealth and its plans for the future of the area. Environmental movement lawyers (ÇHA; *Çevre Hareketi Avukatları*) interpreted this decision as a "*complete massacre*" (*Cumhuriyet*, May 17, 1997). Later, three major economic crises Turkey experienced in less than a decade – in 1994, 1999, and then in 2001 – provided grounds for the state to react to the success story and activism in Aliağa, in typical hegemonic counter-movement fashion. That is, response measures changed the rules of the game – the institutional and legal frameworks – step by step. The first major crisis of the neoliberal era in 1994 was relatively mild, and its ability to dismantle the opposition was rather limited. In contrast, the 2001 crisis constituted a key turning point as it helped discredit the pre-existing policy regime, helped grow the neoliberal wing of the Turkish economic bureaucracy, and both undermined and marginalized the statist opposition to large-scale privatization and deregulation reforms (Öniş, 2004, 2011). Intensified clashes between the national armed forces and Kurdish insurgents in Southeastern Turkey pushed ethnic struggles onto Turkey's main political agenda, which further weakened the environmental opposition.[2]

Indeed, Turkey experienced a major economic liberalization and privatization boom in the aftermath of the 2001 crisis. The country's transformation into a neoliberal economy gained momentum with the privatization of state-owned critical infrastructures. One of the most concrete outcomes of this process has been the shift in the energy sector, where the majority of energy production and transmission passed from the public sector to the private. Many previously state-owned enterprises were privatized in the aftermath of the 2001 crisis – most notably state-owned assets such as the oil company, Petrol Ofisi, in 2002, and the TÜPRAŞ oil refinery in 2005. On the energy front, the government also embarked on a large privatization effort by signing 49-year leases with private firms, granting them usage rights over small rivers and coal mines and enabling them to build and operate hydro and coal power plants (Harris and Işlar, 2013). Unfortunately, implementation of the neoliberal reforms in Turkey was associated with a weakening of the bureaucracy of the state apparatus, arguably with costly consequences. Forgotten for about a decade, the plans for increased coal-fired power capacity in Aliağa resurfaced in the aftermath of this period.

Back to the future in Aliağa (2005–2016)

> *When we meet an investor, they say they will invest but they do not want to crawl at the gates of Ankara [bureaucracy]. So, as the government, we want to solve these issues regarding the permit processes and offer the projects to the investors in a boneless bite, so to speak.*
>
> —Former Minister of Energy, Berat Albayrak (April 21, 2016)[3]

On April 21, 2016, just one day before the official signing ceremony of the Paris Agreement in New York, the Minister of Energy Berat Albayrak (also, the son-in-law of President Erdoğan) met with journalists in Ankara to talk about Turkey's energy strategy. He promised that the *"bureaucratic obstacles"* blocking capital investment in thermal power plants would be removed and these investments would be presented to the investors on a silver plate, or as he puts it, as *"boneless bites."* One such obstacle was the already-decapitated EIA permit process – not the uncertainty of the future of coal in the afterwards of the Paris Agreement as it is in the world. Thus, it is quite telling that after signing the Paris Agreement in New York the next day, the Turkish Minister of Environment Fatma Güldemet

Sarı rushed to the opening of a major lignite-fired power plant in Adana together with President Erdoğan.[4]

Less than a month later, a new mass demonstration took place in Yeni Foça, in the vicinity of Aliağa. Around 2,000 demonstrators consisting of local citizen groups, political parties, national NGOs, and mayors descended on Yeni Foça on May 15, 2016, this time following the call by Initiative Against Fossil Fuels (FYKI, *Fosil Yakıt Karşıtı İnisiyatif*). The demonstration was organized by a local community group in tandem with local authorities but also had a significant back-up by professional environmental groups such as 350.org, European Climate Foundation, and TEMA Foundation (*Türkiye Erozyonla Mücadele Ağaçlandırma ve Doğal Varlıkları Koruma Vakfı* – Turkish Foundation for Combating Soil Erosion, for Reforestation and the Protection of Natural Habitats). The visibility of this event in the run-up to the demonstration proved to be a major instance of revival and remembrance of the memories of earlier anti-coal struggles in Aliağa, with the local movement even appearing in an international documentary film titled *Disobedience*.[5] We contend that such continuities between the mobilizations in 1990 and 2016 help highlight the dynamic of action and reaction between the authoritarian neoliberalism (Tansel, 2018) unleashed by Erdoğan's government and the environmental activists fighting back at local and national scales under transforming political and economic contexts.

During the last decade, strong incentives such as exemption from environmental legislation, highly lucrative subsidy schemes, and generous treasury guarantees were provided for the domestic coal investments, whose return on investments now looked questionable, considering the shift in the global outlook on climate issues after the Paris Agreement. This has however not stopped imported-coal investments, which drive Turkey's worrying current account deficit (Cardoso and Turhan, 2018). Being aware of this trajectory, the government made an amendment to the Energy Market Law in June 2016, which delivered dispatch priority and a purchase guarantee for the electricity generated by power plants using domestic lignite. This policy was intended mainly to keep power companies that have bought existing state-owned coal-fired power plants solvent and to convince the private sector to invest in new lignite power plant projects (Çiftçi et al., 2016). As a result of these neoliberal policies, the share of the privately owned installed electricity capacity – once below the publicly owned capacity – now constitutes more than 75 percent of the total installed capacity in the country.[6] Coal-fired power plants and associated conflicts have been at the forefront of this shift (Arsel et al., 2015).

All this economic and political transformation at the national level also took its toll on the Aliağa region. The once state-led industrialization in the region is now conducted solely by the private sector (in a rather blatant way). Both the state-owned petrochemical industry and the oil refinery were finally privatized in May 2008 despite lengthy protests by the labor union Petrol-İş (once also active in the anti-coal movement) and the ongoing court process against the privatization. After buying PETKİM, SOCAR[7] also bought a whole peninsula in Aliağa, which had "*14 plants, 8 common facilities, power plant, waste treatment plant and a naval port*" (*Cumhuriyet*, May 30, 2018). Clearly, SOCAR's primary aim was not to acquire the now-aging machinery and plants of PETKİM but to get their hands on the valuable land in the area to transform the region into a "*strategic enterprise zone*" (Levent, 2018). Their intentions were demonstrated by one of their first decisions to build an oil refinery and a new 672MW coal-fired power plant on the peninsula.[8] While the plans for another 800MW power plant were shelved due to cancellation of the EIA report by the local court, new coal-fired power capacity of 350MW by İZDEMİR (Izmir Steel and Iron Corp.) emerged in 2009. Despite local resistance, this latter power plant was eventually built and started operating in 2014 while court cases were still ongoing.[9]

Aliağa: *plus ça change, plus c'est la même chose?*

As we were finalizing this chapter, we received news that ministerial approval of EIA reports for two coal-fired power plant projects in Aliağa (one owned by İZDEMİR and the other by SOCAR) were overturned by the local court for the fourth time – annulling the investments on the basis of lack of cumulative impact assessment (diken.com.tr, 2019). Today, carcinogenic risks due to lifetime exposure to volatile organic compounds (half of which may be attributed to petrochemical industries, see Civan et al., 2015) in the Aliağa region are "*substantially higher than the acceptable level*" (Dumanoğlu et al., 2014). In the context of these developments, recounting the story of the past and present anti-coal struggles in Aliağa is helpful for understanding both the socio-economic and political transformation of Turkish society as well as the broader histories of rise and fall of environmentalism in Turkey during the last three decades. In a certain sense, it goes to show how a relatively closed country in its early stages of industrialization opened up to the world and went from a state-led, import-substituting economy into a liberal one first, and then a neoliberal one, while politically oscillating from an authoritarian regime to a relatively liberal one and then back to authoritarianism. This cautionary tale can also be read as the story of erosion of rule of law in emerging, semi-periphery countries in the past 30 years (Saatçioğlu, 2016). In a sense, Aliağa mobilizations in the 1990s set the environmental protest precedent by supplementing its mass mobilization with a strong legal advocacy and therefore provided a coherent paradigm around "environmental rights as human rights," "rule of law," and "democracy" and a repertoire of action for the coming environmental movements such as the Bergama anti-mine struggle (Özen and Özen, 2018). The emergence of the pro-bono legal group, ÇHA from the Aliağa anti-coal struggle has furthermore shaped the next two decades of environmental movements in Turkey. Praised with a newspaper coverage aptly titled "*Name: Lawyer, Surname: Environmentalist*" (*Cumhuriyet*, June 7, 1995), this group eventually led to the formation of a nationally coordinated group, ÇEHAV (Lawyers of Environmental and Ecologist Movements, *Çevre ve Ekoloji Hareketi Avukatları*).[10] Needless to say, the rise of the anti-coal movement in Aliağa, in former mayor Nihat Dirim's words, meant more than solely defense of the local environment:

> *The residues of the coup d'état on September 12th, 1980 were being slowly washed away. [The coup] not only crushed those on the left but also many other parts of the society, a great oppression and fear haunted [the people]. [. . .] Here, we started a social movement with the leadership of municipalities but it went beyond that. The community embraced it and found a space in which it could express itself.*
> (Interview on April 25, 2017)

This micro-history of the anti-coal movement in Aliağa also allows us to unravel the continuities, ruptures, and tipping points in the action-reaction continuum between the state and environmental movements in Turkey. In doing so, it also helps better situate the emergence, evolution, and transformation of the environmental movement in Aliağa and beyond. First, it is clear that there is a strong continuity in the environmental movement through actors and in their repertoires of actions – despite significantly altered relations with the state and the legal system. Many of the current activists remember and long for the 1990s events and their tactics shows a resemblance. For example, in several instances, activists tried to re-create the emblematic human chain action against the coal-fired power plants. In that regard, the environmental activism in the region is still nostalgically reactive, rather than proactive. Second, regarding continuity in the state's policy, it is seen that the neoliberal ambitions of the 1990s are still here albeit now more hostile and unchained from legal hurdles, with the erosion of rule of law in the country. Third, the 1990s discourse that energy scarcity in the country should be immediately addressed is still the dominant *leitmotiv* in 2010s with a twist of "authentic and national energy" (Erensü, 2018).

Despite this aforementioned oscillation between victory and defeat, the emergence of networked grassroots environmental groups offers a glimpse of hope. Following the Gezi Park protest episode in 2013, which culminated in the formation of different neighborhood forums (Uğur-Çınar and Gündüz-Arabacı, 2018; Özdüzen, 2019) across Turkey, local residents upset with being the backyard of the ever-expanding Aliağa industrial area, formed *Yeni Foça Forum* to go beyond a single-issue movement. This new organization, the result of an important reflection process that looked both to the past and to the future, has both produced a discursive and material transformation in the region by claiming to "defend life" (*yaşamı savunmak*) beyond the polluting fossil fuel projects. Despite the limitations due to its rather small member base, the movement's active and openly political stance against polluting investments and active engagement with all other actors has given them leverage to amplify their message. *Yeni Foça Forum* today is an active constituent of the regional platform EGEÇEP (Aegean Environment and Culture Platform, *Ege Çevre ve Kültür Platformu*) and of the national platform Ecology Union (*Ekoloji Birliği*) and has since formed numerous national and international alliances. International NGOs and their national associates (i.e., 350.org and Climate Action Network Europe to name some) are increasingly interested in the region to amplify their messages of climate justice by using Aliağa as an important flashpoint of struggle and advocacy. Pro-bono judicial activism of lawyers from ÇEHAV has also been an important milestone for the Yeni Foça Forum to break away with the financial constraints of due legal processes.

As Knudsen (2016: 322) also concludes in his study of environmental opposition against energy investments in Turkey, informal organization of environmental movements in the country provides opportunities to organize quickly and flexibly "without actually having to comply with any legal requirements and confining procedures." While this flexibility allows them to fly under the radar of the state, thereby rendering environmentalism among the most effective critiques of neoliberal developmentalism (Arsel, 2012), it also gives them the possibility of eventually coming back to fight through multiple and renewed alliances. Like other developing countries, energy has particularly been an important field for environmental movements in the country since it helped politicize environmental movements in Turkey in the aftermath of the 1980 coup – as it allowed politically active individuals and organizations to join forces with local authorities, labor unions, and professional chambers with no previous environmental activism history (Adem, 2005). The anti-coal struggle in Aliağa has particularly been instrumental and arguably pioneered the "legal turn" of environmental activism in Turkey, holding the state accountable for environmental injustices through multi-faceted efforts (Ibid.: 77). Nonetheless, the winning card in Aliağa was the joint effort of "the streets with the parliament, the legal fight with the political fight" in building an ecologist and internationalist narrative (Şahin, 2010).

Kadirbeyoğlu et al. (2017) reiterates that the Turkish state's largely uncompromising position today pushes environmental organizations to make strategic choices with subsequent implications. In a similar fashion, the neoliberalization of Turkey's energy regime and its transformation under Erdoğan's rule proved that not only the state had a comeback as a strong player in the past decade and a half (through synchronizing political, legal, and economic relations with the party-state's preferences) but, also, now it is unafraid to use "heavy-handed legal and extra-legal tools" (Erensü, 2018). Ultimately, confronting an authoritarian neoliberal state unafraid to use coercion and vilification tactics where rule of law does not exist anymore requires environmental justice movements to be propositional as much as oppositional using different and multi-faceted repertoires of contention (Temper et al., 2018a). This, we argue, echoes the calls for resistance-centered perspectives on socio-ecological transformation (Temper et al., 2018b). At this junction, one important question is whether the political pendulum will oscillate back to democratic principles, grounded in the rule of law after the inevitable dissolution of the current authoritarianism. And if this does occur, the question of how the environmental movement will take part in this transformation has no clear-cut answers. Nonetheless, it is clear that local agency has and will continue to have influence over national policies. At the end of the day, it is this agency that will define what type of afterlives environmental movements could have after victories and defeats, towards the political possibilities of common, sustainable, and just futures.

Acknowledgments

We would like to thank Onur Inal and Marco Armiero for their comments that helped us to improve the chapter. We also thank Ayse Ceren Sarı, Billur Biriz, Melis Çelik, and Tanay Özatalar for their research assistance and Aaron Vansintjan for proofreading. We gratefully acknowledge the support and hospitality of Yeni Foça Forum and FOÇEP. This chapter contributes to ACKnowl-EJ project [TKN150317115354] under the Transformations to Sustainability (T2S) Programme coordinated by the International Social Science Council (ISSC) and funded by the Swedish International Development Cooperation Agency (Sida).

Notes

1 Led by climate activism group 350.org and involving a wide range of international, national, and local organizations, Break Free 2016 mobilizations supported 20 popular mobilizations across six continents. Aliağa was one of these 20 sites. https://breakfree2016.org/
2 Any environmentally related social concern would, of course, remain marginal compared to the scale of Kurdish conflict in Turkey.
3 "Kılçıksız yatırım imkanı sunacağız" (We will provide opportunities of boneless investments), Sabah Gazetesi, 21/04/2016, www.sabah.com.tr/ekonomi/2016/04/21/kilciksiz-yatirim-imkani-sunacagiz (Accessed: 30/01/2019)
4 "Cumhurbaşkanı Erdoğan Tufanbeyli Termik Santrali'ni açtı" (President Erdoğan opened the Tufanbeyli Thermal Power Plant) – CNN Türk, 24/04/2016, www.cnnturk.com/turkiye/cumhurbaskani-erdogan-tufanbeyli-termik-santralini-acti (Accessed: 30/01/2019)
5 Another anti-coal demonstration took place in Aliağa in May 2012, organized by a diverse set of national, regional, and local civil society actors and political parties. However, the Break Free mobilization in 2016 was made internationally visible; thanks to the communication support provided by European Climate Foundation and 350.org, see http://watchdisobedience.com/ (Accessed: 30/01/2019)
6 TEIAŞ (Turkey Electricity Transmission Company), Electricity Generation & Transmission Statistics of Turkey, www.teias.gov.tr/T%C3%BCrkiyeElektrik%C4%B0statistikleri/istatistik2015/istatistik2015.htm [Accessed 09.03.2017]
7 SOCAR is an Azerbaijian-owned oil company and one of the world's 50 largest oil companies. The company is also the biggest direct foreign investor in Turkey.
8 Similarly, the company built a wind power plant of installed capacity of 51MW on the peninsula between 2014 and 2017, and announced plans for building a thermal power plant. SOCAR even tried to sideline potential local opposition by taking the chiefs (*muhtar*) of the nearby villages on an all-paid trip to Germany to show how similar "clean" power plants operated.
9 "3 kez ÇED raporu iptal edilen İzdemir termik santraline onay verildi", *Evrensel*, 11/12/2018, www.evrensel.net/haber/368130/3-kez-ced-raporu-iptal-edilen-izdemir-termik-santraline-onay-verildi (Accessed: 02/02/2019)
10 Çevre ve Ekoloji Hareketi Avukatları (ÇEHAV), see http://cehav.org/

Bibliography

Adaman, F., Akbulut, B., and Arsel, M., eds. (2017). *Neoliberal Turkey and Its Discontents: Economic Policy and the Environment Under Erdoğan.* London: I. B. Tauris.

Adem, Ç. (2005). Non – State Actors and Environmentalism. In F. Adaman and M. Arsel, eds, *Environmentalism in Turkey: Between Democracy and Development,* 1st ed. Ashgate: Aldershot, pp. 71–86.

Anadol, K. (1991). *Termik Santrallere Hayır.* V Yayınları: Ankara.

Armiero, M. and Sedrez, L. (eds). (2014). *A History of Environmentalism: Local Struggles, Global Histories.* London: Bloomsbury.

Arsel, M. (2012). Environmental Studies in Turkey: Critical Perspectives in a Time of Neo-Liberal Developmentalism. *The Arab World Geographer,* 15(1), pp. 72–81.

Arsel, M., Akbulut, B., and Adaman, F. (2015). Environmentalism of the Malcontent: Anatomy of an Anti-Coal Power Plant Struggle in Turkey. *Journal of Peasant Studies,* 42(2), pp. 371–395.

Cardoso, A. and Turhan, E. (2018). Examining New Geographies of Coal: Dissenting Energyscapes in Colombia and Turkey. *Applied Energy,* 224, pp. 398–408.

Çetin, E., Odabaşı, M., and Seyfioglu, R. (2003). Ambient Volatile Organic Compound (VOC) Concentrations Around a Petrochemical Complex and a Petroleum Refinery. *Science of the Total Environment,* 312(1), pp. 103–112.

Çiftçi, İ., Berke, M. Ö., and Katısöz, Ö. (2016). Yerli kömüre yeni teşviklerin maliyeti görünenden yüksek olacak. *Fortune Türkiye.* [online] Available at: www.fortuneturkey.com/yerli-komure-yeni-tesviklerin-maliyeti-gorunenden-yuksek-olacak-32183#popup [accessed on February 25, 2019].

Civan, M. Y., Elbir, T., Seyfioglu, R., Kuntasal, Ö. O., Bayram, A., Doğan, G., et al. (2015). Spatial and Temporal Variations in Atmospheric VOCs, NO2, SO2, and O3 Concentrations at a Heavily Industrialized Region in Western Turkey, and Assessment of the Carcinogenic Risk Levels of Benzene. *Atmospheric Environment,* 103, pp. 102–113.

Della Porta, D. (2013). Repertoires of Contention. In McAdam, D. ed., *The Wiley-Blackwell Encyclopedia of Social and Political Movements,* 1st ed. Malden: Blackwell, Available at: https://doi.org/10.1002/9780470674871.wbespm178 [accessed on February 25, 2019].

diken.com.tr. (2019). İzmir'de iki termik santralin 'ÇED olumlu' kararı iptal edildi, *Diken.* [online] Available at: www.diken.com.tr/izmirde-iki-termik-santralin-ced-olumlu-karari-iptal-edildi/ [accessed on February 25, 2019].

Dumanoğlu, Y., Kara, M., Altıok, H., Odabaşı, M., Elbir, T., and Bayram, A. (2014). Spatial and Seasonal Variation and Source Apportionment of Volatile Organic Compounds (VOCs) in a Heavily Industrialized Region. *Atmospheric Environment,* 98, pp. 168–178.

Emek, Y. S. (2015). *Havadan . . . Sudan . . . Allah Allah Havayı Suyu Kim Becerdi?* DT Yayınevi: İzmir.

Erensü, S. (2018). Powering Neoliberalization: Energy and Politics in the Making of a New Turkey. *Energy Research & Social Science,* 41, pp. 148–157.

Harris, L. M. and Işlar, M. (2013). Neoliberalism, Nature, and Changing Modalities of Environmental Governance in Contemporary Turkey. In Y. Atasoy, ed., *Global Economic Crisis and the Politics of Diversity.* London; New York: Palgrave Macmillan, pp. 52–78.

Kadirbeyoğlu, Z., Adaman, F., Özkaynak, B., and Paker, H. (2017). The Effectiveness of Environmental Civil Society Organizations: An Integrated Analysis of Organizational Characteristics and Contextual Factors. *VOLUNTAS: International Journal of Voluntary and Nonprofit Organizations,* 28(4), pp. 1717–1741.

Knudsen, S. (2016). Protests Against Energy Projects in Turkey: Environmental Activism Above Politics? *British Journal of Middle Eastern Studies*, 43(3), pp. 302–323.

Lerner, S. (2010). *Sacrifice Zones: The Front Lines of Toxic Chemical Exposure in the United States*. Cambridge, MA: MIT Press.

Levent, S. (2018). SOCAR Demands 'strategic enterprise zone' in İzmir's Aliağa, *Hürriyet Daily News*. [online] Available at: www.hurriyetdailynews.com/opinion/sefer-levent/socar-demands-strategic-enterprise-zone-in-izmirs-aliaga-131229 [accessed on February 25, 2019].

Müezzinoğlu, A. (2000). A Mediation Case for Resolving the Energy and Environment Dispute at Aliağa-İzmir, Turkey. *Environmental Management*, 26(1), pp. 47–57.

Öniş, Z. (2004). Turgut Özal and His Economic Legacy: Turkish Neo-Liberalism in Critical Perspective. *Middle Eastern Studies*, 40(4), pp. 113–134.

———. (2011). Power, Interests and Coalitions: The Political Economy of Mass Privatisation in Turkey. *Third World Quarterly*, 32(4), pp. 707–724.

Orhan, G. (2006). The Politics of Risk Perception in Turkey: Discourse Coalitions in the Case of the Bergama Gold Mine Dispute. *Policy & Politics*, 34(4), pp. 691–710.

Özal, T. (1987). Turkey's Path to Freedom and Prosperity. *The Washington Quarterly*, 10(4), pp. 161–165.

Özdüzen, O. (2019). Spaces of Hope in Authoritarian Turkey: Istanbul's Interconnected Geographies of Post-Occupy Activism. *Political Geography*, 70, pp. 34–43.

Özen, H. and Özen, Ş. (2018). What Comes After Repression? The Hegemonic Contestation in the Gold-Mining Field in Turkey. *Geoforum*, 88, pp. 1–9.

Özkaynak, B., Aydın, C. İ., Ertör-Akyazı, P., and Ertör, I. (2015). The Gezi Park resistance from an Environmental Justice and Social Metabolism Perspective. *Capitalism Nature Socialism*, 26(1), pp. 99–114.

Özlüer, F., Turhan, E., and Erensü, S. (2016). Türkiye'de Ekoloji Mücadelesi Sınıfsal ve Tarihsel Arayüzler Arasında Nereden Nereye? In S. Erensü, E. Turhan, F. Özlüer, and A. C. Gündoğan, eds, *İsyanın ve Umudun Dip Dalgası*, 1st ed. Tekin Yayınevi: Istanbul, pp. 15–42.

Saatçioğlu, B. (2016). De-Europeanisation in Turkey: The Case of the Rule of Law. *South European Society and Politics*, 21(1), pp. 133–146.

Şahin, Ü. (2010). Türkiye'de Çevre ve Ekoloji Hareketleri Üzerine Notlar: Aliağa Zaferinden Vatan Toprağı Söylemine. *Birikim*, 255, pp. 8–14.

———. (2015). Intertwined and Contested. Green Politics and the Environmental Movement in Turkey. *Südosteuropa*, 63, pp. 440–466.

Şahin, Ü. and Mert, A. (2006). Savaş Emek ile Söyleşi. Yeşil Hareket, Emperyalizm ve Ekoloji. *Üç Ekoloji*, 5, pp. 79–111.

Tansel, C. B. (2018). Authoritarian Neoliberalism and Democratic Backsliding in Turkey: Beyond the Narratives of Progress. *South European Society and Politics*, 23(2), pp. 197–217.

Temper, L., Demaria, F., Scheidel, A., Del Bene, D., and Martinez-Alier, J. (2018a). The Global Environmental Justice Atlas (EJAtlas): Ecological Distribution Conflicts as Forces for Sustainability. *Sustainability Science*, 13(3), pp. 573–584.

Temper, L., Walter, M., Rodriguez, I., Kothari, A., and Turhan, E. (2018b). A Perspective on Radical Transformations to Sustainability: Resistances, Movements and Alternatives. *Sustainability Science*, 13(3), pp. 747–764.

TMMOB. (2016). *Türkiye'nin Enerji Görünümü 2016*. [online] Available at: www.mmo.org.tr/sites/default/files/5a810b69dea7107_ek.pdf [accessed on February 2, 2019].

Tonak, E. A. and Akçay, Ü. (2019). Turkey's Economy Since the 1980 Military Coup. In E. Özyürek, G. Özpınar, and E. Altındiş, eds, *Authoritarianism and Resistance in Turkey*, 1st ed. Springer: Heidelberg, pp. 45–50.

Uğur-Çınar, M. and Gündüz-Arabacı, C. (2018). Deliberating in Difficult Times: Lessons from Public Forums in Turkey in the Aftermath of the Gezi Protests. *British Journal of Middle Eastern Studies*. doi: https://doi.org/10.1080/13530194.2018.1491294.

Yalman, G. L. (2009). *Transition to Neoliberalism: The Case of Turkey in the 1980s*. Istanbul: Istanbul Bilgi University Press.

10 Moving stills

The idea of nature in New Turkish Cinema

Ekin Gündüz Özdemirci

In the 1960s, when film theorist Siegfried Kracauer (1960: 304) argued that film has a potential of "virtually making the world our home", helping us "not only appreciate our given material environment but to extend it in all directions", his intention was to discuss the qualities of a realistic film that could attend to the mind and inner life through the experience of material data. He defined this ability of cinema as the "redemption of physical reality" as he thought of film as the only art along with photography "which leaves its raw material more or less intact" and has the capacity to "read the book of nature" (1960: x). Kracauer saw realism in film as a tool to enhance our intuitive perception of the earth, as he criticized the way science and technology conditioned our attitude towards the material world to an abstract thinking, which *separated* humanity from the everyday surroundings.

Kracauer's ideas challenge the anthropocentric view that promotes an idea of nature that can be calculated, dominated and exploited, by strictly committing to reason and science. This perspective corresponds with Max Weber's (1946: 129) conception of "disenchantment of the world" referring to modern conditions that are attached to causality, decline any mystery and reject the intrinsic value of the more-than-human world independent of its usefulness for the human society. In reaction to that, "enchantment" points out a condition where "all kinds of beings can be existentially alive, subjects with their own qualities and agenda" (Curry, 2012: 3), the wildness of nature cannot be controlled by human will, and "the natural world itself can be a source of the wonder" (Landy and Saler, 2009: 8). This condition is not a smooth harmony, but more like a turbulent field in which various and variable materialities collide, congeal, morph and disintegrate (Khan, 2009: 94).

According to Patrick Curry (2012: 3) "enchantment" that partakes of a non-anthropocentric animism, overturns the official boundaries between human/non-human, animate/inanimate and spiritual/material. Kracauer's ideas on capturing reality through film refer to such inquiry for 'enchantment', a cosmic approach to life, which I believe is also a search for cinematographic eco-awareness; thinking of ecology as a concept emphasizing the holistic existence of all environmental agents. Similar to Kracauer, Gilles Deleuze (1997: 171–172) believed that the link between man and the world was broken, and could be restored within a faith that is readdressed to *this* world. He also relied on film's contribution to restore that broken link.

I argue that a certain amount of realist art house films[1] of the last twenty years, in the period that is generally referred as New Turkish Cinema (Suner, 2010; Dönmez-Colin, 2003; Kaim, 2011), create an "intuitive approach" to nature, which refers to an intimate human encounter with the non-human world that brings out a sense of cosmic unity and hints at the recognition of nature's "intrinsic value". Without having themes related to environmental issues, I claim that these films share a common nature-based aesthetic, with rural narratives that help build up a cinematographic eco-awareness.

However, despite the coherence with ecologically sound philosophies[2], these films develop a binary between nature and society, opposed to the ecological discourse that calls for a relational approach. Nature in these films becomes a refuge, "a place of healing, solace and retreat" as Raymond Williams defines the idea of nature that is separated from human activity (1980: 80). While I discuss the possible motives of this paradoxical approach, I also show that this recent tendency in Turkish cinema still represents a significant shift from environmental concerns to ecological understandings. Even if these films adopt a more subjective and depoliticized perspective on the idea of nature, as there is mainly a conditional valuing of nature's spiritual significance, beauty or diversity, this recognition is still different than the instrumental valuing that we observe in early films from between the 1940s and the 1970s, where nature is only appreciated for what it can bring about for human beings, whereas in recent films it is important for *what it is*.[3]

In early Turkish films, socio-realistic rural narratives were based on insider stories that reflect nature as domicile, as a place of economy and labor relations, however through a prominent anthropocentric perspective. Urbanite directors of the New Turkish Cinema transform this social realism into a cosmic realism, by which Anderson and Hausman refer to the view that the universe is or has an aspect that is extra-mental (2012: 45). Francesca Mari argues that cosmic realism is the late descendant of the Romantic and Transcendentalist traditions, where the attempt to discover the deepest truths moves beyond science, into the spiritualized self (2008). Cosmic assessment of life overlaps with the idea of the 'enchanted' world that is material and spiritual at the same time, mysterious, undelimitable and unmasterable (Curry, 2012: 2–3). Frank Luger defines cosmic realism as a worldview that shows the modest place of humans in nature, and nature's proper place in humans, while attracting attention to the argument that the "cosmos is the true measure of all things" (2007: 588). According to this worldview, reality is not limited to human species' perception and understanding of it. That complies with the expression of Great Mystery concerning the universe, which was adopted by many scientists (Taylor, 2009: 220) such as Loren Eiseley, Aldo Leopold or Rachel Carson who said "every mystery solved brings us closer to the threshold of a greater one" (Carson, 1998: 159).

In this chapter I use cosmic realism to define a cinema that places emphasis on the emotional and spiritual aspect of the affinity between human and non-human nature beyond a physical interdependency, and manifest a "natural world that includes but vastly exceeds humanity" (Curry, 2012: 5). While observing the everyday life in its simplicity this cinema builds up an intuitive approach to non-human nature through film image, and interrogates the human perception of reality as it "fluctuates between actual and virtual, sometimes confusing mental and physical time" (Totaro, 1999).

The countryside, which has a distinctive place in Turkish film history, becomes the location where the stories unfold long after its popularity in early films. First examples of environment-related themes in Turkish cinema consisted of socially realistic village stories, and narratives on the tension between the city and the countryside resulting from geographically uneven development. Within a less outspoken everyday reality instead of an open socio-political context, realist art house films of New Turkish Cinema follow personal stories based on a search for belonging, mainly from an outsider perspective of urbanite directors and characters. These films lead to interrogations of current connotations and conflicts around nature in urban space by turning away from the urban area that is either shown, or depicted in dialogues as a *synthetic* "second nature", and by contrasting it with "nature as a given context", a less controlled or untouched ecosystem that is found in the rural area. In this respect I think these rural films are much more related to urban socio-environmental realities than they seem to be and enable a discussion on urban-rural dichotomies in contemporary Turkey.

This chapter examines the rural contexts in recent Turkish films that become a meeting point with an external and non-social nature that is substantially devoid of human intervention, and discusses its relevance with shifting connotations of the urban space. I will place the discussion around the analysis of three films, *Mayıs Sıkıntısı [Clouds of May]* (1999), *Yumurta [Egg]* (2007) and *Koca Dünya [Big Big World]* (2016), because they are based on narratives that are directly related to the issues of non-belonging around urbanite characters that have left the city for rural spots. *Big Big World* follows the story of fugitive orphans, whereas *Egg* and *Clouds of May* deal with an uneasy homecoming. Besides being examples of cinematographic eco-awareness, these films also display a common idea of nature that is more like a refuge or retreat from society. As common in Turkish film history, Istanbul is the urban location in these films, as well as in most recent films with urban settings, and I show that the quality of urban transformations in Istanbul is relevant with the exposition of rural area and the idea of nature in New Turkish Cinema.

The chapter begins with a look at early Turkish films, observing the shift from environmental concerns to ecological understandings in recent films. I then briefly examine the dynamics of urban ecology after the 1990s, focusing on Istanbul to comment on the possible grounds of shifting rural narratives from an outsider urbanite perspective. Finally, I give examples of the cosmic vision and embodied ecological awareness in recent Turkish films by an aesthetic and narrative analysis, and discuss the idea of nature they display in a social context.

Nature as 'environment' in early cinema

Narratives based on an urban-rural dichotomy have been a distinctive character-istic of Turkish cinema since the early years of the newly founded Republic in the 1920s. The main concern of the films was then to contribute to the modernization ideals of the Republic. Urban areas, mainly Istanbul, were praised as the center of modernism against the rural life that was considered the site of conservative tradi-tions to be transformed. The essence of human relations to the natural environ-ment was mostly restricted to anthropocentric contexts related to the supposedly principal indicators of modernization such as industrial development or mechani-zation in agriculture. Since these developments placed progress under state control, the state and modernism were adopted as identical concepts, so building up the image of a modern country was a matter of national identity and the state was over-protective for that matter through a strict policy of censorship in various sectors.

Under such circumstances, the urban locations in the newly emerging Turk-ish cinema reflected ideas of growth and Istanbul was especially portrayed as the center of attraction, while rural settings could only become an idyllic location in escapist melodrama films between the 1940s and 1950s. In order to avoid the accusations of "insulting the Turkish villager", most films of that period turned the rural space into an exotic background (Scognamillo, 1973: 9). *Dark World [Karanlık Dünya]* (1952), a film by Metin Erksan, one of the pioneer directors of social realism in Turkish cinema, and *Land [Toprak]* (1952) by Nedim Otyam were subjected to censorship for accusations directly related to environmental issues (Aytekin, 2015: 321; Köse, 2011: 104). This act of censorship for showing Turkish land as arid and barren uncovers how the government twisted socio-environmental realities and valued natural environment as a manifestation of a strong national identity. The last scenes of *Dark World* reflect the intervention of censorship; the close-up images of modern agricultural machinery are shown in parallel with the images of the new healthcare center built in the village, being indicators of government services in the making of a modern society.

"Village films" of the 1950s were mostly adapted from "village novels", which then became a popular literary genre. Along with the constitution of the Vil-lage Institutes (*Köy Enstitüleri*)[4], a whole new generation of writers started to address socioeconomic problems in rural areas. Between the 1950s and 1970s, novelists such as Fakir Baykurt, Mahmut Makal, Yaşar Kemal, Necati Cumalı and Orhan Kemal developed a socially realistic literary approach to rural themes. Taner Timur (1991: 357–358) defines the core context of the village novels as "a democratic revolution philosophy that opposes all feudal remains and exploita-tion mechanisms, oppression of landlords and loan sharks against poor villag-ers in Turkish rural". Cinema also corresponded to this literary fashion with an increasing number of rural narratives; however, censorship prevented the rapid evolution of social realism in films that reflect the revolutionary philosophy of village novels. Considering the low literacy rate in the first years of the Turkish Republic, it is more likely that cinema was regarded as a more efficient media for reaching a wider public than literature.

After the enactment of a constitution with more libertarian edges in the 1960s, Turkish cinema entered a new phase of social realism between the 1960s and 1970s with films mostly addressing environmental issues around the urban-rural dichotomy. The films such as *The Revenge of the Snakes [Yılanların Öcü]* (1962), *Dry Summer [Susuz Yaz]* (1963), *The Law of the Border [Hudutların Kanunu]* (1967) focused on the struggles of rural people who mix their labor with the natural environment. By incorporating environmental issues such as water or land ownership, these stories encouraged a new kind of political awareness that criticized feudal structures and capitalism. Environmental issues then became a mirror for the imaginary of a productive nation and a politically conscious society that knows how to claim rights within modern relations, against conservative traditions.

The Law of the Border, one of the early socio-realistic films by Lütfi Akad, highlights the solidarity and collective struggle for the rights to education and agricultural production against conservative and destructive traditions in the region, such as obeying landlords and smuggling at the border. Most of the films of this period place female characters at the center and build up a parallel between the patriarchal domination over natural resources and that over women's bodies. In that they become the symbol of resistance, female characters can be considered as part of the stance against the conservative traditions that mainly dominated the rural areas and exacerbated the urban-rural conflict. Metin Erksan, in three of his rural films, *The Revenge of the Snakes*, *Dry Summer* and *The Well [Kuyu]* (1968), respectively defends the independency of land, water and women as life-giving "resources". The discussion on water and land takes shape around the equal share of these "resources" by the people whose livelihood depend on them in rural areas, and women characters who are at first victimized are the ones who find a way to initiate a resistance.

The Trap [Tuzak] (1976) problematizes the intertwined connections between capitalism and environmental pollution in a melodramatic narrative influenced by social realism. The story follows the murder of a mayor by a greedy factory owner who refuses to clean toxic waste and protect the water resources. The film represents environmental pollution as the result of human degradation within capitalist relations, and hints at the importance of local resistance. In 1978, *The Herd [Sürü]*, one of the most important socio-realistic rural films in Turkish film history, also used the close-up images of the modern agricultural machinery like the previously mentioned *Dark World [Karanlık Dünya]* did in 1952; but this time it was not referring to the benefits of modernism, but rather pointing to the depreciation of small farmers and the regression in rural productivity. *The Herd*, which is the tragedy of a nomadic family who has to sell their animals and immigrate to the city, is one example of the rural-urban migration narratives in Turkish cinema that also unfold environmental issues from the 1970s onwards.

In the period of social realism Turkish films finally began to interrogate the charm of Istanbul and observe a chaotic and uncanny city reflecting the effects of increasing immigration from rural areas. Lütfi Akad is behind the most prominent socio-realistic "immigration films" of the 1970s with his trilogy *The Bride [Gelin]* (1973), *The Wedding [Düğün]* (1974) and *Blood Money [Diyet]* (1975), which focused on the migrants in Istanbul. These films portrayed the socio-spatial exclusion of villager immigrants as a result of cultural conflict, lack of workplace safety and illegal housing in shantytowns at the outskirts of the city. Similar to Metin Erksan, Akad approached the problem of internal immigration in the context of women's victimization and placed female characters in the center of the struggle for social rights. These films, observing the effects of migration on traditional values and human relations, also give insights on the transforming spatial ecology of the city, with an attentive portrayal of growing settlements and immersive industrial areas in the background.

The socio-realistic films of the 1960s and 1970s mainly revolve around environmental issues in the context of their impacts on human society. Environmental themes in these filmic narratives are tools for promoting socio-political awakening, which is very much in compliance with the highly politicized society of the time[5].

Ecological understandings in New Turkish Cinema

Following the coup d'état in 1980, Turkish cinema entered into a period of recession until the 1990s. The 1990s represent a rebirth for the Turkish film industry with a whole new generation of educated directors who made their first films in this period. Turkish cinema, after a long time, saw box-office success with popular films, and prestigious awards at international film festivals with art house films. Many films in this period remarkably engage with ecologically sound rural narratives and reveal a more-than-human perspective.

We can consider *Tarzan of Manisa [Manisa Tarzanı]* (1994) as the first feature film that is not restricted to modernist or socialist imaginaries in relation to the natural environment, embracing a more holistic approach to human and non-human nature relations. It is adapted from the real life story of Ahmet Bedevi[6], nicknamed as the "Tarzan of Manisa", who lived from 1899 until 1963 and is known to be the first environmental activist of the Turkish Republic. The film draws attention to the importance of preserving the ecosystem and the consequences of deforestation in the context of the conflict between ecological priorities and the goals of modernization. Besides underlining the significance of local civilian activism, the narrative also embraces an ecological awareness and reflects on the intrinsic value of nature independent from human needs and priorities, in parallel with the life philosophy of Bedevi.

Some films of this period that substantially take place in rural areas, such as *Sellale* (2001), *Sour Apples [Ekşi Elmalar]* (2016) and *Home [Yurt]* (2011) can be considered as "eco-nostalgia films" (Özdemirci and Monani, 2015), referring to Asuman Suner's (2010: 37) definition of Turkish nostalgia films where rural life of the past is positioned as a lost paradise and the "mise en scène is often dominated by aestheticized images of rural landscape". What we observe in these films is a past, or an imagery of it, that often comprises a communal life that has qualities of living in harmony with nature. In these films, personal and social loss are identified in parallel with an environmental damage; it is either the construction of a dam that blocks the waterfall, an abandoned orchard and a local apple species that become extinct, or an idyllic landscape that is destroyed by mining activities that change the course of events or overturns the idealized past.

Didactic comedy films like *Ecotopia [Entelköy Efeköy'e Karşı]* (2012) and *Dairy Philosopher [Mandıra Filozofu]* (2014) embody bioregional ideals of renewing and integrating human and non-human communities. These films can be identified to have an "ecotopian" vision as they tell the stories of going back to nature, rejecting the modern economic and social relations, and building an off-grid and self-sufficient communal or individual lifestyle in rural areas in line with voluntary simplicity. Nostalgia comes in the scene again as "going back to nature" evolves as an act encouraged by a longing to and an appreciation of pre-modern ecological values. But in this case nostalgia does not manifest itself with mourning for a loss and a search for healing through that, it is rather a progressive mobilizer, as the past values become an instructive and guiding resource for the present.

What I define as ecological awareness in New Turkish Cinema emerged in the same period as the civilian environmental activism. Since the first years of the Turkish Republic, environmental conservation has been mostly a state controlled matter. As observed in early Turkish films, human relation to non-human nature was identified in parallel with national modernization ideals, which favored business and industrial interests in the development of environmental policies for decades (VanderLippe, 2011: 216). From the beginning of the 1990s, environmental movements along with other social right movements, such as women's and human rights and peace initiatives, evolved to be an important part of civil society. State policies have also put the environment on the agenda. According to the first Environment Law published in 1983, regulations and precautions to make better use of and preserve the land and natural resources in rural and urban areas had to be in conformity with economic and social development objectives. This appeal has been mostly applied with an anthropocentric impulse that puts economy before ecology. Unstructured and distorted understandings of development and economic growth, primarily related to constructional development in urban areas and energy policies in rural areas, resulted in increasing rural-urban migration and environmental problems, also triggered civilian movements, especially locally driven campaigns.[7]

Distancing from nature in the urban space

Rural narratives of New Turkish Cinema that unfold the urbanite's search for belonging in relation with a natural environment that becomes an isolated refuge from society overlap with urban films of this period that reveal the image of a chaotic urban space. Most films with urban settings take place in Istanbul and manifest the destructive and depressing effects of urban transformations, dealing with feelings of non-belonging and alienation.

As in the past, Istanbul continued to be the preferred urban setting for Turkish cinema. Besides the rich cultural heritage of the city, economic circumstances are also influential in this accustomed location choice. As with most industries, the film industry has been centralized in Istanbul since its inception. Even though the rural-urban migration began in the mid-1950s, a large-scale decrease in rural population occurred especially after the 1980s (Yılmaz, 2015: 163), and being a pole of attraction for local and foreign investments and import-substitution policies, Istanbul was the "dominant center of production, exploitation and accumulation" (Soja, 1980: 221) and had the highest migration rates in this period.

The impacts of urban sprawl and shrinkage of green spaces in Istanbul have become especially visible since the 1990s, including developments such as the construction of a second bridge on the Bosporus (TMMOB Chamber of City Planners Istanbul Branch, 2010: 20).[8] Urban films identify this aggressive concretion in the city with nearly dystopic scenes where green spaces are totally excluded from landscapes. Narratives on the parallelism between the degradation and alienation in human relations and the synthetic urban ecology (*Block C*, 1994; *Somersault in a Coffin*, 1996; *Innocence*, 1997; *Pandora's Box*, 2008; *Istanbul Tales*, 2005; *My Father's Wings*, 2016), displacements due to urban transformation (*Song of My Mother*, 2014) or ongoing urban development (*Dream*, 2016) reflect destructive transformations in the city.

The immigration films of the 1970s portrayed the industrial settlements and factories around the urban space, however urban structures are altered in recent urban films as they mainly highlight the building trade and places like shopping malls that emphasize the service sector. Deindustrialization of the urban economy, and policies that envisioned Istanbul as a global center for financial and cultural industries, led to large-scale development projects[9] such as business towers, shopping malls, luxury hotels or clearance of squatter areas for fast pace lucrative redevelopment (Karaman, 2008: 518–521). Gentrification has become a general policy in the re-organization of urban space with the construction sector and housing market being the driving forces of the economy, particularly since the early 2000s (Harmanşah, 2014; İslam, 2010; Karaman, 2008; Ergun, 2004). Gentrification imposes a certain way of living in line with consumer culture, which turned some inhabitants into excluded, thus alienated minorities. Exclusion takes different forms; it can result in pressure over cultural, political, racial, economical differences, as well as ecological understandings. Besides being an issue of human and social rights, I think that thwarting the right to ecological living in the city is also a form of exclusion. These exclusionary practices can be witnessed in the imposed urban transformations that take away the opportunity to practice an alternative ecological way of urban living, as the market interests govern the decisions on natural resources, public spaces and green areas.

The process that led to Gezi Park resistance is a recent example that mirrors the rising frustration among some inhabitants, mainly "educated and young urban crowds" (Harmanşah, 2014: 123), towards imposed decisions to transform public spaces and green areas. There were many triggering factors in this process, but ecological alterations in Istanbul and its reflections in social life paved the way. Environmentally controversial projects such as the construction of a third bridge across the Bosporus, a third airport on the remaining forestland around the city, the announcement of Canal Istanbul, a canal project that would connect the Black Sea and the Marmara Sea and the plans of a new urban development on the surrounding lands successively came to the fore. At the same time, systematic interventions in the Taksim area, which had histori-cally been the intellectual center of alternative culture, dramatically changed the socio-spatial structure of the neighborhood. Symbolic cultural places such as historical movie theaters, patisseries, cafés and bookstores had to shut down due to increase in rents or development projects. The historic buildings that changed property and were transformed into shopping malls, and the related boom in property values, display how gentrification shaped the re-organization of space and culture.[10]

While the authorities talked about creating a 'new Istanbul' around the Canal project, they were already remaking the current urban space, and the Taksim area is only one example of this large-scale transformation. The project of rec-reating the historic artillery barracks that used to exist from 1780 to 1940 on what is now Gezi Park, thus transforming the park into a commerce and cultural center, was announced at the same time. The ongoing replacement of cultural memory, accompanied by the constant exclusion of green spaces, culminated in the Gezi Park development project. That gave rise to a nationwide resistance for the "defense of urban historical heritage or the collectively used city spaces, which are always deeply saturated with social memory, and senses of belonging" (Harmanşah, 2014: 123).

Urban transformations that took place in Istanbul, especially after the 1990s, represent an example of what Lefebvre identifies as "capitalism's success in achieving growth by occupying space and by producing a space" (1976: 21). The production of space in Istanbul resulted in forming a *synthetic* 'second nature'. Lefebvre's take on the notion of 'second nature' stresses the importance of see-ing and creating the cities as built environments in compliance with nature and socio-spatial diversity, rather than restricting the idea of nature to an external pure state separated from human activities. New Turkish Cinema embodies reflec-tions of this socio-environmental agenda in Istanbul, both in urban and rural films. The films that will be examined in the next section take a stand against the 'disenchanted' world by achieving a more-than-human perspective around sto-ries of urbanites searching for belonging in relation to the natural environment. However in return, they repeat the pattern and fall into escapism by limiting well-reflected ecological awareness with an idea of nature that is 'enchanted' only in isolation from society.

External nature in rural films

The directors of the Turkish art house cinema, Nuri Bilge Ceylan, Reha Erdem and Semih Kaplanoğlu, relate to the natural environment without directly including environmental issues in their narratives. The stories in *Clouds of May*, *Egg* and *Big Big World* follow alienated or troubled characters that gravitate from their urban environments to the rural dimensions. The countryside is portrayed as a space where one can communicate with an undisturbed nature, which unfolds memories, the subconscious or points to an innate reciprocity between human and non-human.

The realistic rural films differ in aesthetic qualities such as the long take, like landscape shots that are one to seven minutes long, or spontaneous observations of everyday life. The revelation of the natural environment is also determinant in terms of narrative as it displays a reflection of human conditions, but not reduced to human meanings or agendas. This is where an intuitive approach comes in and adds a cosmic dimension to realism with an attempt to create a non-hierarchical bond between human and non-human nature, to capture the 'wildness' of nature and recognize its intrinsic value.

The prominent Soviet filmmaker and film theorist Andrei Tarkovksy is one of the pioneers of such cinematic style mentioned above. The spontaneous rhythms of natural phenomena such as the stream of a river, rainfall, fog or snow "underscore and often dictate" (Totaro, 1992: 23–24) the rhythm of his films, both in aesthetic and narrative terms. What he called "the life of nature" associates with the 'wildness' of the natural world that "includes but vastly exceeds humanity" (Curry, 2012: 5). This sustains the dramatic development, merging non-human and human existence through memory, imagination and consciousness. Tarkovsky built a parallel between his film approach and the qualities of Japanese haiku art, which is known as poetry based on a minimalistic lyrical phenomenology of ordinary aspects of life. Just like reading a haiku, the observation of life through film image was for Tarkovsky an invitation to be absorbed into the depths of the cosmos where there is no bottom and no top (Tarkovsky, 1987: 106–107). Similar to Kracauer's idea on film's capacity to "read the book of nature", for Tarkovsky what the viewer sees on the screen "is not limited to its visual depiction, but is a pointer to something stretching out beyond the frame and to infinity; a pointer to life" (1987: 118). A cosmic search through realism in film encourages the director to disclaim a predictive power over nature and the non-linear mind. This unpredictive approach is the true potential of film to attain a cinematographic eco-awareness or create a "cinematic non-anthropocentrism" as Justin Remes (2016: 225) stated.

Clouds of May (1999)

Clouds of May is the second feature-length film by internationally acclaimed director Nuri Bilge Ceylan. Along with his short film *Cocoon [Koza]* (1995) and first feature-length film *The Small Town [Kasaba]* (1998), it is a part of his so-called rural trilogy[11]. The film is about a film director, Muzaffer, returning to his childhood hometown from Istanbul to make a movie about his family, especially about his father. His father, Emin, is bent on saving the small forest he cultivates on his property from confiscation by the authorities.

The story mainly develops in Emin's land and the small forest. The film is based on observing and appreciating him and his intimate relation with the forest, both from his son Muzaffer's and the director Nuri Bilge Ceylan's urbanite perspectives. The actor who plays Emin is Ceylan's own father and the land also belongs to him in real life, so he substantially plays himself in the film. The film reflects Emin's intimate connection to nature in long shots of his labor on the land and with photographic scenes showing him as an organic part of the unified space. Close-up shots, and camera transitions between him and the plants in his land, especially the old oak trees, intensify a sense of connection and unity. Different than the portrayal of labor relations with the natural environment in early rural films, non-human nature is not captured only as surrounding elements, but occupies a prominent presence on the screen in its own rhythm.

The film places Emin in a completely different reality compared to the rest of the characters. His values and priorities contrast with the urban and modern values that identify Muzaffer and his film shooting agenda. The small forest he cultivates near his property, which he desperately tries to protect from confiscation by the authorities, is presented as a counter-world. Muzaffer aestheticizes the place and his father's connection to natural environment through experiments with his camera, while Ceylan as the director also introduces poetic images of his father working and living on his land. In the film Ceylan uses some images from his short film *Cocoon* as if Muzaffer took them. In these scenes we see moments of the daily life of his parents in the land, along with the elements of the natural environment such as wind, thunder, rain clouds and sky. Classical music in the background merges with sounds of nature, and slow camera transitions from the father's close-up face to the tree leaves and sky create a sense of cosmic intertwining. The images sometimes get blurred in transition thus putting into question the supposed human and non-human boundaries. The film ends with Emin eating an apple he collected from one of the trees, and then falling asleep as he leans against a tree in the forest. As he sleeps, the camera turns to the trees in front of him; the sun rises slowly from among the trees with the chirping of the birds and the whispering of the wind. The screen is filled with sunlight and the sounds of the natural environment continue as the credits roll. Altogether these scenes capture an alternative impression of time in nature that is indifferent to the bustle of human society.

Egg (2007)

We follow the story of an alienated urbanite in *Egg*, whose experience of a nature in isolation also becomes prominent. Yusuf who lives in Istanbul returns to his childhood hometown upon his mother's death. He has to stay longer unwillingly to fulfill his mother's wishes true by performing a sacrifice. In a short scene where we have an idea about Yusuf's life in Istanbul, unsettlement is foregrounded. He does not have a home, sleeps in the second-hand bookstore where he also works.

Egg is the first film of the award-winning trilogy of director Semih Kaplanoğlu and it is followed by *Milk [Süt]* (2008) and *Honey [Bal]* (2010). These films are based on the life story of a character named Yusuf in different stories with common features. In the last film *Honey*, we see that Yusuf's journey as a child takes shape with his father dying in the forest while searching for honey. Yusuf takes shelter in a tree hollow when he learns his father's death. This loss represents the end of his childhood and his detachment from nature at the same time. In *Egg*, we see Yusuf's only published book of poetry entitled 'Honey'. In an interview, Kaplanoğlu explains that he considers honey as the "soul of the forest". Yusuf's return to his hometown turns into a spiritual search pursued in an untainted nature, which signifies a search for integrity and belonging in relation to his past and memories.

The natural environment becomes an alternative ground of self-confrontation that draws Yusuf into a spiritual journey of dreams and memories. After his mother's funeral, the camera slowly follows him, stopping on the tree branches flapping in the wind in a nearly one-minute-long scene as he walks towards the forest from the cemetery. He falls asleep there and wakes up from an odd dream where we see him breaking a quail egg.

The almost six-minutes-long opening scene that shows Yusuf's mother walking towards the forest, merging with the shadow of the trees as she disappears from sight, foreshadows her death. This lengthy scene is like a poetic statement on death, nature and cosmic unity. We learn later that his mother has kept her deceased beloved ones alive through the plants that she grew. She gave the name of her dead husband, brother or sister to different flowers and talked to them until her death. The film establishes and reinforces this cosmic approach as Yusuf relates to the natural environment that almost takes the place of his mother, whom he couldn't communicate with until her death, and for which he feels guilty. In a scene where he surrenders himself in an untainted natural setting, we witness his emotions for the first time. While he is on his way back to Istanbul he stops his car on the roadside and walks into a bush. As the sun goes down a giant sheepdog appears suddenly and pulls him down by jumping on him. Yusuf stays immobile where he fell in order to prevent another attack of the dog and remains for hours, in the middle of the bush, surrounded by the darkness. As the dog stands over him, growling occasionally, Yusuf suddenly starts weeping loudly. He surrenders himself to the shifting balance of power between human and non-human, and in a way experiences a catharsis. As he wakes up by an olive tree in the woods he decides not to go back to Istanbul.

Big Big World (2016)

We observe a pristine nature turning into a ground of spiritual journey and a counter-world even more prominently in the award-winning director Reha Erdem's *Big Big World* (Figure 10.1). The film tells the story of two orphan teenagers, Ali and Zuhal, who grew up together with a bond as strong as between brother and sister. Ali commits a crime to save Zuhal from abuse and an arranged marriage, and they run away from Istanbul to find shelter in the woods.

Erdem also portrays a nature devoid of human activity, and places it as an alternative world to the urban space that is defined by hierarchy and patriarchal authority. The film prominently separates the harsh reality of the city by incorporating its cruelty and exclusion, and the mystery of the woods where human

Figure 10.1 Movie poster, *Big Big World*

memories and unconscious unfold through intuitive encounters with non-human nature. Similar to Kaplanoglu, Erdem also uses the image of sleeping in nature's bosom as a metaphor of spiritual journey and the pursuit of integrity in many of his films[12] including *Big Big World*. In a couple of scenes we see the sleeping teenagers as if they were an integral part of the natural environment. Like in most of Erdem's previous films, a strong presence of plants, animals and natural elements like wind or water occupy a significant amount of screen time. The film shows detailed images from the everyday rhythm of non-human nature, highlights its wildness beyond human society, while building up a psychophysical connection between human and non-human through its main characters. Erdem's main characters are always children, women, old people or outsiders who are somehow abused by patriarchal society. Their intuitive connection with non-human nature, especially with animals, is a repeated pattern in his films.

In *Big Big World* Zuhal spends some time alone in the woods, while Ali works in the nearest small town. She communicates with animals and plants and encounters an old woman. As the old woman dies, Zuhal remains for some time near her corpse. This experience turns into a spiritual connection and a compelling healing process. Death and life are intertwined; Zuhal covers the body with dry leaves and feels that the dead body is as alive as the natural environment. She lies down and sleeps on the soil, her face covered with the dead woman's scarf, or she swims in the lake where her still body interacts with the fish; she holds the dead woman's hand or touches tree trunks with a meditative attitude. Towards the end, Zuhal becomes physically ill and mentally disturbed by memories of abuse. She starts appealing to the goat she always comes across as her father, and suddenly her body responds with a bleeding that looks like a miscarriage. Scared of losing her, Ali takes the girl to a hospital in the city. While he is waiting outside and hiding from the police, the same goat appears to him. Overwhelmed, Ali also starts seeing the animal as his father, and cries out for help. Anthropomorphism is only restricted to the characters' perception, the film does not reveal that and prevents any sense of identification for the audiences. The detailed and long shots of animals are reminiscent of the aesthetics of nature films in the voyeuristic sense, but there is an objective view of the natural world, without any emotional or violent framing.

Nature as a counter-world

More than the rural space itself, the external, non-social nature that the characters of these films encounter in the countryside is at the core of these narratives, and I believe this adds to the binary approach that the films constitute between nature and the society. Nature becomes a counter-world that is devoid of societal relations, and represents a more embracing and genuine structure opposite to the idea that the films create about urbanity, which is identified with notions such as unsettlement, degeneration, alienation or authority. On the other hand, rural life is more like a "degenerate mixture" (Davoudi and Stead, 2002: 269) of urban and rural qualities. It is a space of underdevelopment and related problems.

In *Clouds of May* through the story of Saffet, an unemployed young villager, we witness the lack of access to jobs and little economic diversification in a rural area. Saffet feels trapped in the countryside and overestimates the opportunities in Istanbul. The film opens up with his story as he learns that, after several unsuccessful attempts, he has once again failed to be admitted at the university. Out of desperation, he starts working at a factory but soon loses his job. Ceylan's next film *Distant [Uzak]* (2002) portrays the sequel of Saffet's story, where the villager finds himself once again excluded from, and alienated within, the big city.

On the other hand, Emin's struggle to save the forest hints at the destructive politics of development on forestlands in rural areas. Throughout the film we watch Emin as a part of the forest and land, but his belonging is under threat because the authorities try to confiscate the forest. He believes that he can prevent the division of the land by defending the integrity of the ecosystem in the area. The legislative system, and the contradictions in its application, is a notable backstory in the narrative. Towards the end of the film Emin discovers the marks made by the authorities on some of the trees he had cultivated. As he wanders in the forest, the sound of chainsaws in the background, reminding us of the 'disenchantment' of society, contrasts with the sounds of the natural environment and mirrors his unease and the threat of destruction.

Nuri Bilge Ceylan observes the urban-rural dichotomy in a personal context by examining his own urban perception of his father's relation to the land. Nature is a separated, restricted matter to be preserved against a human-induced danger, something fragile in *Clouds of May*. Mixing labor with an intuitive approach to the natural environment is something that belongs to the older generation, close to extinct and under threat, while the younger generation is longing for urban life and its facilities that failed to be adapted to rural life.

Big Big World also addresses in parallel form the poverty and despair related to corruption in human relations both in the city and in the countryside. Urban scenes are set in the poor neighborhoods at the outskirts of Istanbul, in a zone surrounded by industrial estates and concrete buildings. Night shots and darker colors accentuate the atmosphere of misery. Ali lives in a dorm-like ratty apartment that he shares with three other teens, and he is mistreated in the motor mechanic workshop where he works. Zuhal lives with her abusive foster family who does not let her go out or see her brother.

The small town that is close to the forest where they take refuge shares similar patterns with the city. Ali is attracted to a woman working in a traveling fair, but gets swindled by her and loses all his money. In the forest, the characters eventually encounter uncanny strangers from the town, who secretly wander around, observe them or harm their bike. Nature is a refuge, a temporal shelter from society in *Big Big World*, and human presence does not refer to cooperation but to unease and threat. The film separates the human society from nature in a way that displays the corrupted societal relations as unnatural.

In *Egg* Kaplanoğlu has a more naïve and idealized portrayal of the rural than the other two directors. Through characters such as the warm-hearted young girl Ayla, a distant relative who had been living with Yusuf's mother for years, the rural is portrayed as a place of goodwill and cooperation that gradually softens Yusuf's alienation and draws him in. Unlike other films, *Egg* does not support its portrayal of nature-society dualism with a critique of human society, but underlines the conflicting urban and rural customs in support of the latter.

From Yusuf's point of view rural life is reflected as a dysfunctional, ancient site where time has kind of stopped in the beginning of *Egg*. The film portrays the contrast between cognitive time, in relation to Yusuf's memories that come back through his presence in an external nature, and the time he spends in his rural town that pushes him into involuntary social relations with the funeral process. While an external and non-social nature becomes a place of solace and self-confrontation, we see Yusuf in conflict with rural life and his memories of it. When he meets his ex-girlfriend, she remembers him saying that he would never live anywhere else but this town. However, Yusuf has a different memory, according to him he always hated that place. But the film ends with his decision not to go back to Istanbul; instead he goes back to his family home where Ayla is waiting. The last scene shows them having breakfast together.

Appraised rural customs are restricted to traditional everyday practices and human relations that remain exotic without any critical approach, such as Ayla buying milk from the milkman who is coming to the house, Yusuf searching for egg in the coop, an old man spinning yarn with an ancient waving tool or the mutual assistance between town people etc. Kaplanoğlu has a more conservative approach to the urban-rural dichotomy. His portrayal of the rural world leans towards a "restorative nostalgia", which Svetlana Boym defines as preserving an idealized stable image of the past, in a way, turning the past values into a holy shrine (2009: 87). The film creates a conflict between Yusuf's alienated modern habits and idealized traditional indicators in a healing "back-to-the-roots" story, that eventually holds onto the questionable idea of an "authentic rural" that also embodies an "authentic" untainted nature.

Conclusion

The impossibility of home within the context of an urban-rural dichotomy and nature-society dualism in New Turkish Cinema is not distinct from the story of separation shaping the natural and social environment in contemporary Turkey. Narratives evolving around a search for belonging in 'enchanted' nature that is isolated from the 'disenchantment' of society are like a yearning for "rediscovery at-homeness in the world" (Landy and Saler, 2009: 11). Dismissing deep-rooted ecological instincts is considered a cause of grief, despair and anxiety (Smith, 2010: 3), which can be a common ground for many people living in urban-industrial societies, but especially in cities like Istanbul where massive urban development has been the manifestation of the city's retreat from nature over more than a decade. Such reorganization of urban space transforms natural environments as well as the ecology of societies, which results in the alteration of memory that uproots people from their social and cultural identity.

Recent rural films in Turkish cinema reflect a longing for comfort, security and emotional attachment against a state of alienation, displacement and frustration. Building up a cosmic approach, these films offer a non-hierarchical understanding of life, appreciation of diversity and intuitivism, which are in opposition to discourses of domination, separation and rationalism that turn the world into inanimate and quantitative matter. The films highlight an inherent bond between human and non-human with an intuitive perception of 'enchanted' nature, and provide a much-needed insight. However, they also reveal an alienation from society and its living spaces, and favor a search for meaning in the alternative reality of intuitive senses. Psychologist Martin Jordan argues that a naïve mindset that sees nature as a perfect, benevolent parent is a form of escapism, possibly a sign that someone is "less in love with nature than out of love with society" (Smith, 2010: 5). The idea of nature as a counter-world isolated from the perils of human society embodies a naïve escapist tone, and mirrors the impulse of the urbanite against the current socio-environmental agenda in Turkey, which produces feelings of frustration and exclusion.

Despite the cosmic approach and intuitive perception of human and non-human relations, creating a binary between nature and society places the ecological awareness in a dilemma. Overcoming this requires the adaptation of such intuitivism to the inevitable shaping of nature by human activities, both in urban and rural spaces. As Williams (1980: 79) says, "a state of nature could be a reactionary idea, against change, or a reforming idea, against what was seen as decadence". Besides valuing nature as sanctuary of psychophysical catharsis and mindfulness, cinematographic eco-awareness should also embrace insights on social actions in line with this valuing, such as making the rural areas alive again by building up ethical labor relations between humans and non-humans, or creating a regenerative 'second nature' in urban areas. Underlying this inseparability of nature, culture and society is the real potential of film to "read the book of nature".

Notes

1 In industrial terms art house is used to define mostly independently made films, aimed at a niche audience. In film studies the definition is expanded to describe a film's distinctive visual and narrative qualities. As Steve Neale phrases, "art films tend to be marked by a stress on visual style . . . by a suppression of action in the Hollywood sense, by a consequent stress on character rather than plot and by an interiorisation of dramatic conflict" (1981: 13). In art cinema "cause-effect logic is attenuated in favor of slow, reflective causation" with "long takes, real time exposition and a characteristic slowness of pace" that adds to realism, or with "a montage that mixes up story order" with "violations of real time to express the psychological experience of characters and their actions of reactions" (Betz, 2009: 11)

2 For concepts of ecocentrism and biocentrism, see Taylor, 1986 and Curry, 2011.

3 For further information on the objective and subjective intrinsic value of nature, see Vilkka, 1997 and Skolimowski, 1986.

4 The Village Institutes, a national education initiative to train teachers, health officers, technicians and members of professions who would serve the needs of the villages, got started in 1936 with 21 centers around the country and were legalized in 1940. This

was a large-scale educational mobilization that aimed to improve rural life with vocational trainings as well as cultural activities in art, literature and sports. The Village Institutes were officially closed down in 1954 under the conservative government of the period, with claims of communist propaganda rumored by the feudal sector that considered the increasing rural consciousness as a threat against traditional power dynamics. (For more information, see Kocabaş, 2017.)

5 The decade of the 1960s saw the rise of leftist movements in Turkey, similar to the tendencies in many countries. Political debates between left- and right-wing parties turned into more radical and violent conflicts in the street among various factions formed around universities and civilian initiatives, including armed attacks and bombings that killed many people on a daily basis. These events resulted in two coup d'états in March 12, 1971, and September 12, 1980, which resulted in death penalties, controversial trials and eventually the regression of leftist movements and the expansion of a free market economy (For more information, see Zürcher, 2017, Ulus, 2010; Keyder, 1987.)

6 Ahmet Bedevi or Ahmeddin Carlak was born in 1899 in Samarra under the rule of the Ottoman Empire. He served in the Turkish Army during the Turkish War of Independence and settled in Manisa in western Turkey after the war ended. He dedicated himself to the reforestation of war-ruined mountains and lands in the town. As he lived in a small cabin at Mount Sipylus without any modern furniture, wore only shorts in all seasons, shared his small income from gardening work at the municipality with poor people and appealed to all plants and trees he cultivated as "my children", local people embraced him and called him Tarzan of Manisa after the famous fictional character Tarzan created by Edgar Rice Burroughs in 1912. Bedevi died in 1963, and every year a memorial for his death anniversary is linked to World Environment Day celebrations on June 5 (Şahin, 2015: 444–445).

7 The decade of the 1990s saw the first large-scale civilian environmental movements in Turkey, also the pioneer victory of Turkish environmental activism. It was against a thermal power plant project in the industrial town of Aliağa near Izmir. The campaigns that started in 1989 with the support of municipalities, local NGOs and people became stronger with the formation of a 50km-long human chain from Aliağa to Izmir and resulted in the cancellation of the project by the government. Although thermal power plants continue to be a problem in the environmental agenda of Turkey, local resistances and victories also continue growing.

8 Green space per person in Istanbul is currently far below the amounts in other big cities like London with 33%, Rome 34.8%, Madrid 35% or Vienna 44.5%; Istanbul Metropolitan Municipality declared this amount as 6% for Istanbul, but according to the Foresters Association of Turkey it is only around 1% (TMMOB Chamber of Environmental Engineers, 2017).

9 The volume of foreign trade increased threefold between 2002 and 2007 and Istanbul had a share of 60% in this increase. As an example of market-oriented urban development, between 2004 and 2009, the number of shopping malls has increased around sevenfold (İslam, 2010: 60).

10 Historical buildings on Istiklal Street in the Beyoğlu district from the 19th century, such as Cercle d'Orient, Sin-Em Han or Narmanlı Han, changed hands and transformed into shopping malls and commerce centers. These buildings used to be homes to historical movie theaters such as Emek Cinema, symbolic places of urban heritage such as İnci Patisserie, and places of social gatherings such as small theaters, publishing houses or cultural and political institutions, which have either been displaced or vanished.

11 His famous film *Distant* (*Uzak*, 2002) is sometimes considered one of his rural narratives. Even though the story mainly takes place in Istanbul, it is a great portrayal of urban-rural conflict as a town dweller migrates to a big city to seek a job. He

temporarily stays with his well-educated relative who has long been alienated by his rural roots.

12 *Times and Winds* (*Beş Vakit*, 2006) revolves around the story of three children, their rite of passage, conflicts with the adult world and how patriarchal relations shape their growing up in a village in western Turkey. Erdem uses mystical sleeping scenes as a leitmotif. Each time the children are forced by the adult world and exposed to maltreatment or injustice, we see them sleeping under a tree, in the bushes or on the rocks as if they are organic parts of these environments. Erdem underlines a division between patriarchal human society and nature through these symbolic scenes. Many eco-philosophers, especially the deep ecologists, draw a parallelism between a child's intuitive perception of nature and the idea of ecological self to become convergent with the interest of the rest of life (Naess, 1986: 19; Taylor, 2001: 180). Children are the only characters who don't yet belong to the human-made patriarchal system, they rather act and feel like a part of nature.

References

Anderson, D. R. and Hausman, C. R. (2012). *Conversations on Peirce: Reals and Ideals*. New York: Fordham University Press.

Aytekin, P. E. (2015). Köy Gerçekliği Bağlamında Türk Sinemasında Edebiyat Etkisi: Lütfi Ömer Akad Sineması. *Gazi Üniversitesi İletişim Kuram ve Araştırma Dergisi*, 41, pp. 314–330.

Betz, M. (2009). *Beyond the Subtitle: Remapping European Art Cinema*. Minneapolis, MN: University of Minnesota Press.

Boym, S. (2009). *Nostaljinin Geleceği*, trans. F. B. Aydar. Istanbul: Metis.

Carson, R. (1998). *Lost Woods: The Discovered Writing of Rachel Carson*, Linda Lear, ed. Boston: Beacon Press.

Curry, P. (2011). *Ecological Ethics: An Introduction*. Cambridge: Polity Press.

———. (2012). Enchantment and Modernity. A Draft of a Paper in: *Philosophy, Activism, Nature*, 12, pp. 76–89. [online] Available at: www.patrickcurry.co.uk/papers.htm.

Davoudi, S. and Stead, D. (2002). Urban-Rural Relationships: An Introduction and Brief History. *Built Environment*, 28(4), pp. 268–277.

Deleuze, G. (1997). *Cinema 2 The Time Image*. Minneapolis: University of Minnesota Press.

Dönmez-Colin, G. (2003). New Turkish Cinema – Individual Tales of Common Concerns. *Asian Cinema*, 14(1), pp. 138–145.

Ergun, N. (2004). Gentrification in Istanbul. *Cities*, 21(5), pp. 391–405.

Harmanşah, Ö. (2014). Urban Utopias and How They Fell Apart: The Political Ecology of Gezi Parkı. In U. Özkırımlı, ed., *The Making of a Protest Movement in Turkey: #occupygezi*, 1st ed. London: Palgrave Pivot, pp. 121–133.

İslam, T. (2010). Current Urban Discourse, Urban Transformation and Gentrification in Istanbul. *Architectural Design. Special Issue: Turkey: At the Threshold*, 80(1), pp. 58–63.

Kaim, A. A. (2011). New Turkish Cinema – Some Remarks on the Homesickness of the Turkish Soul. *Cinej Cinema Journal*, Special Issue 1, pp. 99–106.

Karaman, O. (2008). Urban Pulse-(Re)Making Space for Globalization in Istanbul. *Urban Geography*, 29(6), pp. 518–525.

Keyder, Ç. (1987). *State and Class in Turkey: A Study in Capitalist Development*. London: Verso.

Khan, G. (2009). Agency, Nature and Emergent Properties: An Interview with Jane Bennett. *Contemporary Political Theory*, 8, pp. 90–105.

Kocabaş, A. (2017). Village Institutes in Turkey and Cooperative Learning. *European Journal of Education Studies*, 3(3), pp. 47–64.

Köse, Ö. (2011). Türk Sinemasında Sansür. M.A. Istanbul: Istanbul University.

Kracauer, S. (1960). *Theory of Film: The Redemption of Physical Reality*. New York: Oxford University Press.

Landy, J. and Saler, M. (2009). *The Re-Enchantment of the World: Secular Magic in a Rational Age*. Stanford: Stanford University Press.

Lefebvre, H. (1976). *The Survival of Capitalism*. New York: St. Martin's Press.

Luger, F. (2007). *Farbensinn: A Collection of Scientific Essays*. Montreal: Self published.

Mari, F. (2008). Cosmic Realism. *The New Republic*. [online] Available at: https://newrepublic.com/article/62425/cosmic-realism.

Naess, A. (1986). Deep Ecology in Good Conceptual Health. *The Trumpeter*, 3(4), pp. 18–22.

Neale, S. (1981). Art Cinema as Institution. *Screen*, 22(1), pp. 11–40.

Özdemirci, E. G. and Monani, S. (2015). Eco-nostalgia in Popular Turkish Cinema. In S. Rust, S. Monani, and S. Cubitt, eds, *Ecomedia: Key Issues*, 1st ed. London: Routledge.

Remes, J. (2016). The Sleeping Spectator: Non-human Aesthetics in Abbas Kiarostami's Five: Dedicated to Ozu. In T. de Luca and N. B. Jorge, eds, *Slow Cinema*, 1st ed. Edinburgh: Edinburgh University Press, pp. 231–242.

Şahin, Ü. (2015). Intertwined and Contested. Green Politics and the Environmental Movement in Turkey. *Südosteuropa: Journal of Politics and Society*, 63(3), pp. 440–466.

Scognamillo, G. (1973). Türk Sinemasında Köy Filmleri. *Yedinci Sanat*, 4, pp. 8–14.

Skolimowski, H. (1986). In Defense of Ecophilosophy and of Intrinsic Value: A Call for Conceptual Clarity. *The Trumpeter*, 3(4), pp. 9–12.

Smith, Daniel B. (2010). Is There an Ecological Unconscious? *The New York Times*. [online] Available at: www.nytimes.com/2010/01/31/magazine/31ecopsych-t.html?pagewanted=print

Soja, E. W. (1980). The Socio-Spatial Dialectic. *Annals of the Association of American Geographers*, 70(2), pp. 207–225.

Suner, A. (2010). *New Turkish Cinema: Belonging, Identity and Memory*. New York: I. B. Tauris.

Tarkovsky, A. (1987). *Sculpting in Time*. Austin: University of Texas Press.

Taylor, B. (2001). Earth and Nature-Based Spirituality (Part 1): From Deep Ecology to Radical Environmentalism. *Religion*, 31(2), pp. 175–193.

———. (2009). *Dark Green Religion*. Berkeley; Los Angeles; London: University of California Press.

Taylor, P. W. (1986). *Respect for Nature: A Theory of Environmental Ethics*. Princeton: Princeton University Press.

Timur, T. (1991). *Osmanlı-Türk Romanında Tarih, Toplum ve Kimlik*. Istanbul: Afa.

TMMOB Chamber of City Planners Istanbul Branch. (2010). *3. Köprü Projesi Değerlendirme Raporu*. Istanbul: Chamber of City Planners Istanbul Branch. [online] Available at: www.mimdap.org/images/dosya/spoist_3.koprurapor.pdf.

TMMOB Chamber of Environmental Engineers. (2017). *İstanbul Çevre Durum Raporu*. Istanbul: Istanbul Chamber of Environmental Engineers. [online] Available at: www.cmo.org.tr/genel/bizden_detay.php?kod=96492&tipi=67&sube=2.

Totaro, D. (1992). Time and the Film Aesthetics of Andrei Tarkovsky. *Canadian Journal of Film Studies*, 2(1), pp. 21–30.

———. (1999). Gilles Deleuze's Bergsonian Film Project: Part 2. *Off Screen*, 3(3). [online] Available at: https://offscreen.com/view/bergson2

Ulus, Ö. M. (2010). *The Army and the Radical Left in Turkey.* London: I. B. Tauris.

VanderLippe, J. (2011). The Statist Environment: Gazi Orman Çiftliği and the Kemalist Modernization Project. In S. Oppermann, U. Özdağ, N. Özkan, and S. Slovic, eds, *The Future of Ecocriticism: New Horizons,* 1st ed. Newcastle upon Tyne: Cambridge Scholars Publishing.

Vilkka, L. (1997). *The Intrinsic Value of Nature.* Amsterdam; Atlanta, GA: Rodopi.

Weber, M. (1946). Science as a Vocation. In H. H. Gerth and C. Wright Mills, eds, *Max Weber: Essays in Sociology,* 1st ed. Oxford: Oxford University Press, pp. 209–226.

Williams, R. (1980). *Problems in Materialism and Culture.* London: Verso.

Yılmaz, M. (2015). Türkiye'de Kırsal Nüfusun Değişimi ve İllere Göre Dağılımı (1980–2012). *Eastern Geographical Review,* 20(33), pp. 161–187.

Zürcher, E. J. (2017). *Turkey: A Modern History.* London: I. B. Tauris.

Films Cited

Anlat İstanbul [Istanbul Tales]. (2005). [film] Istanbul: Ümit Ünal et al.

Annemin Şarkısı [Song of My Mother]. (2014). [film] Istanbul: Erol Mintaş.

Babamın Kanatları [My Father's Wings]. (2016). [film] Istanbul: Kıvanç Sezer.

Bal [Honey]. (2010). [film] Istanbul: Semih Kaplanoğlu.

Beş Vakit [Times and Winds]. (2006). [film] Istanbul: Reha Erdem.

Block C. (1994). [film] Istanbul: Zeki Demirkubuz.

Diyet [Blood Money]. (1975). [film] Istanbul: Lütfi Akad.

Rüya (Dream). (2016). [film] Istanbul: Derviş Zaim.

Düğün [The Wedding]. (1974). [film] Istanbul: Lütfi Akad.

Ekşi Elmalar [Sour Apples]. (2016). [film] Istanbul: Yılmaz Erdoğan.

Entelköy Efeköy'e Karşı [Ecotopia]. (2012). [film] Istanbul: Yüksel Aksu.

Gelin [The Bride]. (1973). [film] Istanbul: Lütfi Akad.

The Herd [Sürü]. (1978). [film] Istanbul: Zeki Ökten.

Hudutların Kanunu [The Law of the Border]. (1967) [film] Istanbul: Lütfi Akad.

Karanlık Dünya [Dark World]. (1952). [film] Istanbul: Metin Erksan.

Kasaba [The Small Town]. (1998). [film] Istanbul: Nuri Bilge Ceylan.

Koca Dünya [Big Big World]. (2016). [film] Istanbul: Reha Erdem.

Koza [Cocoon]. (1995). [film] Istanbul: Nuri Bilge Ceylan.

Kuyu [The Well]. (1968). [film] Istanbul: Metin Erksan.

Mandıra Filozofu [Dairy Philosopher]. (2014). [film] Istanbul: Müfit Can Saçıntı.

Manisa Tarzanı [Tarzan of Manisa]. (1994). [film] Istanbul: Orhan Oğuz.

Masumiyet [Innocence]. (1997). [film] Istanbul: Zeki Demirkubuz.

Mayıs Sıkıntısı [Clouds of May]. (1999). [film] Istanbul: Nuri Bilge Ceylan.

Pandora'nın Kutusu [Pandora's Box]. (2008). [film] Istanbul: Yeşim Ustaoğlu.

Sellale. (2001). [film] Istanbul: Semir Aslanyürek.

Susuz Yaz [Dry Summer]. (1963). [film] Istanbul: Metin Erksan.

Süt [Milk]. (2008). [film] Istanbul: Semih Kaplanoğlu.

Tabutta Rövaşata [Somersault in a Coffin]. (1996). [film] Istanbul: Derviş Zaim.

Toprak [Land]. (1952). [film] Istanbul: Nedim Otyam.

Tuzak [The Trap]. (1976). [film] Istanbul: Atıf Yılmaz.

Uzak [Distant]. (2002). [film] Istanbul: Nuri Bilge Ceylan.

Yılanların Öcü [The Revenge of the Snakes]. (1962). [film] Istanbul: Metin Erksan.

Yumurta [Egg]. (2007). [film] Istanbul: Semih Kaplanoğlu.

Yurt (Home). (2011). [film] Istanbul: Muzaffer Özdemir.

Epilogue

Üstün Bilgen-Reinart

In the spring of 2018, unprecedented flash floods alarmed the citizens of Ankara, the Turkish capital. Floodwaters dragged away dozens of cars, hundreds of businesses suffered damages and countless people were wounded. Later in November 2018, this time in the upscale resort town Bodrum on the Aegean coast, torrential rains preceded by two waterspouts and hail led to flashfloods. In a luxury shopping mall, visitors waded waist high in water. Local and central state authorities acknowledged the disasters as 'God's will' but hardly anyone in power or in mainstream media uttered the words 'climate change' or even pointed at the connection between frenzied extractive projects, fossil fuel burning power plants, urban transformation, de-forestation and climate change. Climate change, once a distant risk, is now a catastrophe in progress.

As a university student in Canada in the 1970s, a journalist through the 1980s and the 1990s, and a writer and activist in Turkey from the late 1990s on, I have reported on the growing encroachment of global capitalism on the natural and social world. I have witnessed ecological struggles emerging as calls for justice and isolated instances of protest in Canada. In the early 1990s, as a news reporter, I visited and reported on aboriginal communities disenfranchised by the Churchill River diversion for hydroelectric dams at Southern Indian Lake in Manitoba, Canada. The indigenous people were not resisting or protesting loudly then, although their lakes and the fish they ate were contaminated with mercury and their spruce trees had sunk into the water because of soil erosion in melting permafrost. Their suffering was all but invisible at that time. Their voices were muted against the backdrop of the heady expansion of global neoliberalism.

Things have changed in the last three decades. Neoliberalism has been transforming the relationship between nature and societies so radically that ecological collapse and social disenfranchisement have given rise to resistance movements that now erupt (and often face violent retaliation) frequently in many local communities across the world. Pipeline projects to transport oil extracted from the tar sands of Northern Alberta to refineries in Illinois and Texas or to a distribution center in Oklahoma through traditional aboriginal lands have provoked one of the most dramatic ecological resistance movements in history at Standing Rock, North Dakota. Similarly, on the Pacific coast of Canada, the Wet'suwet'en First Nation that has never signed treaties with the government of Canada, and thus has legal jurisdiction over their unceded territories, has vowed to resist gas pipe-lines at any cost, and set up road blocks. Across South America, indigenous com-munities are opposing mines, dams and agri-businesses. Indigenous people are now in the forefront of the global resistance against social transformations caused by ecological destruction.

In the last decade of the 20th century, the resistance of villagers to a gold mine in Bergama was among the first '*indigenous*' environmental movements in Turkey. I recorded the oral history of those villagers. Long before the fertile fields around the village of Ovacık were turned into barren pits of toxic, cyanide-con-taminated debris, I was with the villagers on many occasions when hundreds of them marched to the gates of the gold mine, with women in the front and men, stripped to the waist, bringing up the rear, calling out "*Halkız, Haklıyız, Kazanacağız* (We are the people, Our cause is just, We shall win!)."[1]

They did win – for a while. They won wide support from civil society. Pro-fessional chambers, unions, environmental groups and many ordinary citizens in urban settings supported and even joined their struggle. As Yaşın points out in her chapter on the anti-mining resistance in this book, the resistance to the gold mine in Bergama exemplified "local articulations of a global environmental justice movement." The resistance succeeded in delaying the operation of the Bergama-Ovacık mine for more than a decade, and thus postponed the onslaught of mining companies that already held more than 600 drilling licenses in Tur-key. It seriously disturbed the Turkish government and the trans-national mining companies they were allied with. They managed to break the resistance with treachery. A story invented by the Turkish secret service, alleging the oppo-nents of the mine were German spies, alarmed the villagers and their supporters (Akdemir, 2011).

Once the Bergama resistance was suppressed with "spy trials" at the *Devlet Güvenlik Mahkemesi* (now defunct State Security Court), where all the accused were acquitted, the damage done by the smear campaign remained. Meanwhile, other gold mines began to set up shop all over the country. In the last two dec-ades, Turkey has given 360,000 mining licenses for stone quarries, coal, nickel, copper and gold mines. The laws and regulations protecting olive groves, agri-cultural lands, coasts and historic sites were amended in 2004, 2010 and 2015 to allow mining companies greater access. Today, 13 gold mines are active in Turkey, and many more investors hold drilling licenses.

One of those active gold mines, at Kışladağ in western Turkey, has the dubi-ous distinction of being the largest gold mine in Europe. Canadian Eldorado Gold Corp.'s open-pit, heap leaching mine sprawls over 157 km^2 in the prov-ince of Uşak. It started operations in 2006 amidst reports of cyanide poison-ing and deformed farm animals (Sakaryalı, 2011) and still continues to date, having increased its capacity three times in 2011, 2014 and 2016 respectively. Efemçukuru, located 23 km from the third-biggest metropolitan city Izmir, also belongs to a subsidiary of the same Eldorado Gold Corp. Situated in the water catchment area of Tahtalı Dam that provides 40 percent of Izmir's drinking water, Efemçukuru mine threatens Izmir's water supply with arsenic contamination, at a time when climate change is predicted to exacerbate and intensify droughts in the region.

In *The Iliad*, Homer referred to the mountains in the Biga peninsula in North-western Anatolia as "Ida of a thousand springs". Considered as the lungs of the Marmara region, the forests on Ida Mountains are rich with important sources of water, dozens of endemic plants and great bio-diversity. The mountains are also culturally, mythologically and archaeologically priceless. It is not an exaggeration to say Ida Mountains are life-giving and life-sustaining. Yet, drilling for gold on Ida Mountains started in 2007, and the ecologists in the region report 16 companies hold gold-mining licenses to extract the metal on 34 points, using approximately 400,000 tons of cyanide. İsmail Pehlivan, the President of the Chamber of Agricultural Engineers, says drilling operations at the towns of Ayvacık, Bayramiç, Çan, Lapseki, Bize, Yeniköy and Edremit are already causing irreparable damage to agricultural landscapes.

As discussed throughout this book, Turkey has striven to develop and control its natural resources since the establishment of the Republic in the early 20th century. From those nation-building days on, public works and construction projects have been considered tools of prosperity, and their ecological costs largely disregarded. Today, unregulated resource extraction serving the global mining industry, and industrial projects and grandiose construction plans profiting pro-government capital have taken on a frenzied pace. Environmental Impact Assessments (EIA), still required by law for many polluting investments, have become all but sham. The country's energy system relies heavily on imported fossil fuels such as petroleum, natural gas and coal with the share of fossil fuels in energy production being as high as 85.5 percent in 2016. In 16 years of AKP's rule, Turkey has constructed hundreds of hydro-electrical plants and dozens of coal-fired thermal power plants despite strong opposition from villagers, scientists and ecologists.

The nuclear adventure

Although Turkey was among the countries severely affected by the Chernobyl disaster of 1986, the Turkish government still seems determined to go ahead with plans to build its first nuclear power plant. It has been four decades since a site at Mersin's Akkuyu district was licensed in 1976 for the first nuclear power plant in the country. After decades of inaction (due to economic as well as political circumstances), the Turkish government eventually signed a bilateral agreement with the Russian Federation in May 2010 for the construction and operation of a 4800MW nuclear power plant at Akkuyu, effectively handing Akkuyu over to Russia for 15 years to produce electricity above market rates.

Shortly after this agreement, the Fukushima disaster in Japan on March 11, 2011, dampened the government's enthusiasm only for a short time. Not only did the Turkish government continue to develop the nuclear energy project at Akkuyu, but in 2006, it also signed an agreement with a consortium of French and Japanese companies for a second nuclear power plant in the city of Sinop at the Black Sea coast. Thousands of people from all over the country protested, including fisherfolk who paraded at sea in their boats with anti-nuclear banners. Despite these protests, 225,000 trees were cut down on the İnceburun peninsula of Sinop where this power plant was planned. Strong opposition continues, but the anti-nuclear movement is increasingly strangled by arbitrary bans of protests as exemplified by the April 2018 protest ban enacted for the first time since mass protests started in 2006. In December 2018, Japanese newspapers reported that the country would be withdrawing from the project because of the rising cost of the nuclear plant due to the looming economic crisis. Turkish authorities have so far made no announcement confirming or denying this.

Mega dams

At a time when the ecological devastation caused by large dams all over the world is beginning to be acknowledged and the benefits being questioned, Turkey is going ahead with a particularly destructive phase of its ambitious GAP (*Güneydoğu Anadolu Projesi*, Southeastern Anatolia Project). GAP was undertaken in the 1970s with plans to build 22 dams and 19 hydro-power plants to harness more than 50 billion cubic meters of water that flow annually through the Tigris and the Euphrates rivers, representing 28 percent of the country's waters.

Ilısu Dam – a grandiose hydraulic project submerging the town of Hasankeyf and its 12,000 years of heritage, displacing up to 70,000 people and destroying an important eco-system with dozens of species of birds, reptiles and fish in the Tigris Valley – is the most controversial of the dams in the GAP. Ilısu claims to be the largest dam of its kind in the world: concrete-face rock-fill dam, 135 meters high, with six units each with installed capacity to produce 200 megawatts of electricity, 4.12 billion kilowatt hours per year at a total economic cost of 1.7 billion USD. The destruction of the town of Hasankeyf, which is already happening with water detention underway in Ilısu Dam, will also deal a severe blow to Kurdish history and culture, very likely one of the purposes of the project. Turkey's unilateral claims to the right to open or close the floodwaters of the Tigris and the Euphrates rivers will certainly cause tension with Iraq and Syria. As Rıdvan Ayhan, the spokesperson of Hasankeyf Initiative (*Hasankeyf'i Yaşatma Girişimi*), confirms 199 communities and 289 archaeological sites will the submerged by the dam. Today, historic structures such as a Seljuki tomb, baths dating from the Artuq reign in the 12th century, and a medieval mosque with double minarets are already being removed piece by piece to be re-built and displayed elsewhere.

The third airport in Istanbul

On October 29, 2018, the Turkish President Recep Tayyip Erdoğan was photographed leading a parade of golf carts with visiting dignitaries riding amusement-park style at the inauguration of "one of the world's largest airports", the third International Airport in Istanbul to the north of the city on the shores of the Black Sea. This grandiose facility covering an area of 76,500,000 m^2 is a catastrophe by any measure. Geologists have warned the swampy, muddy and artificially filled ground is not solid and the filling might give way. Meteorological conditions such as intense fog, harsh winds and ice on the Black Sea Coast threaten flight security (Northern Forests Defense, 2015).

But even more importantly, the area chosen for the airport was covered by the only standing forest in the vicinity of Istanbul, with endemic flora and fauna, and several sources of water including the Alibey Lake, Pirinçci Dam and Terkos Lake that supplies a large part of the drinking water for this mega metropolis. In all, 70 wetlands were dried to cover the lush land with concrete. Some lakes will be drained into the Black Sea, changing sea levels. At a time of runaway climate change and increased demand for water, the destruction of the last and only green space near Istanbul is nothing less than a criminal act. There will be a tremendous loss of several eco-systems with an inevitable deterioration of air quality for Istanbul. In all, 2.5 million trees including Anatolian chestnuts, oaks, eastern spruces, linden, ashes and hornbeams were cut to make way for President Erdogan's pet mega-project.

Already, in December 2018, after heavy rains, the construction site of the giant airport was flooded submerging vehicles and construction equipment. The ecological cost of Turkey's neoliberal frenzy now threatens not only the humans but also non-humans (such as wild boars escaping to the city looking for food) and all sources of life: air, soil, water and food supply. Today, ecological struggle is a struggle for life itself. Hence, those resisting the lawless plunder of nature and heritage in Turkey call themselves "*yaşam savunucuları* (defenders of life)".

Environmental struggles: then and now

Under neoliberal authoritarianism in Turkey, at a time of frenzied, ill-researched, ill-conceived mega projects, both spontaneous, local resistance and organized resistance[2] offer some hope. In Turkey, as in many other countries in the world, during the 1980s resistance to destructive projects was scattered and muted. When the now-privatized Yatağan coal-fired thermal power plant began to operate in 1985, spewing out mercury and ash over villages and olive groves, local groups politely appealed to the national energy company (*Türkiye Elektrik İletim A.Ş*; TEAŞ), the Governor of Muğla and the Minister of the Environment, to get it closed. Nothing happened. The local people then went to court and obtained expert reports against the coal power plant. They still lost their case and eventually appealed to the European Court of Human Rights (ECHR). ECHR ruled in July 2005 that the plant should be closed, but instead of shutting down, the plant operators increased capacity and enlarged the coal mining site, encroaching on olive groves. Today, the Yatağan power plant is among the least productive plants in Turkey, yet it is still operating after having displaced the village of Yeşilbağcılar. Some of the villagers were re-settled in hastily constructed apartments but many others were scattered and impoverished. Incidents of cancer have greatly increased in the region. Climate Action Network (CAN) Europe's recent study concluded that three power plants in the region are responsible for the massive air pollution with severe consequences: Yatağan, Yeniköy and Kemerköy coal-fired power plants were responsible for an estimated 45,000 early deaths between 1983 and 2017.[3]

In the 21st century, the social and economic impacts of neoliberalism have greatly exacerbated the socioeconomic and environmental problems in Turkey as in the rest of the world. Ecological degradation is accompanied with social exclusion, severance of agricultural ties to the land and food insecurity. Privatization, deregulation and state-sanctioned urgent land expropriations have caused widespread poverty and inequalities in local communities. In response, environmental resistance movements have changed character, becoming more widespread and desperate. Villagers across regions of Turkey now hold 24-hour vigils and set up road blocks to prevent construction of roads, power plants, mines and dams. They sometimes succeed in delaying (if not hindering) projects, much like indigenous peoples all over the world.

Such spontaneous resistance in the country has produced an unexpected and ironic result: rural women have become active in the frontlines of environmental justice movements. The ecological and social transformations disenfranchising their villages and in many cases destroying their livelihood have freed women and propelled them into more active roles as subjects. In Bergama, several women described to me how they had slipped out of submissive roles inside their homes to take part in demonstrations against the gold mine on short notice, without even consulting their husbands or fathers. A similar prominence of women has emerged in Artvin, during the Cerattepe resistance to gold mining.

Throughout the second decade of the 21st century, when resistance to run-off hydro-electric power in the Black Sea region gained steam, once again it was women who chased company officials out of their villages. Influential women made headlines, such as Neşe Öztürk from the village of Yaylacılar in Black Sea town of Rize who told a BBC reporter in 2013 that she and other village women were strong in the resistance because they worked so hard at home that they knew the struggle for water was a struggle for life. Similarly, in places like Çamlıhemşin, Gerze and Yırca (and more recently in Kızılcaköy in western Turkey, fighting back geothermal power investments), women actively hold vigils night and day to block machinery often at the expense of rough treatment at the hands of the gendarmes. This has become a pattern.

Important as they are, it is difficult for such 'indigenous' resistance movements to become effective agents in challenging the hegemony of neoliberalism as long as they remain localized, fragmented and limited. It is important to remember that local groups resisting an extractive project are not just confronting a company, they are actually facing national and global economic and political powers. That is why hope lies in the possibility of mobilizing national and international allies. If local groups succeed in strengthening alliances with others in civil society, sustaining cohesion and broadening their struggles on a foundation of political insights, they may actually become agents of change in challenging neoliberalism.

Legal battles: then and now

Since the 1990s, when ecological resistance in Turkey came of age with the opposition to the gold mine in Bergama, pro-bono lawyers have played a major role in
the movement, challenging destructive projects in the courts. During the resistance to the Bergama gold mine, lawyers in the movement took on the State and
transnational companies, won some important rulings on behalf of public good,
such as the Council of State decision of 1997 against the Bergama gold mine,
and thus they have etched the struggle on record. Since then, consecutive governments have disregarded such progressive court decisions and systematically
amended legislation to prevent legal obstacles to their preferred industrial and
mining projects. Today, Turkey is no longer a country where the rule of law is
paramount.

In today's Turkey, ecological battles may be won in local courts, but higher
courts are likely to reverse those decisions. In the case of Bergama, a decision in
2018 has overruled the progressive, pro-public court decision of 1997, thereby
contradicting the court's own legitimacy. A general sense of despair is thus
growing towards the now dubious legal process in the defense of environmental
integrity, notably after the constitutional change in 2017. The Justice and Development Party (AKP; *Adalet ve Kalkınma Partisi*) government has significantly
raised the costs of launching a court case and of fact-finding missions by experts.
The courts now use random excuses to refuse admission to some complainants. In
short, the legal struggle in defense of the environment has become difficult. Since
court decisions are often not even respected, environmental advocacy groups
(such as EGEÇEP) are giving up on lengthy and costly court battles.

Yet it is important not to abandon the legal struggles. Lawyers and citizens
defending life still launch court cases and succeed in obtaining court decisions
against many life-threatening projects. One of those lawyers, Arif Ali Cangı from
Izmir Bar Association, says that in spite of legal irregularities, corporate secrecy,
extralegal approval of EIAs, hasty regulations that replace protective legislation
and increasing lawlessness, legal battles still remain an important field of resistance. Legal struggles on environmental matters familiarize courts and judges with
burning ecological issues and reassure communities that they can have recourse
to justice. Furthermore, they also yield important expert reports from fact-finding
missions, which in turn serve as crucial institutional memory. Well-researched,
impartial expert reports (where available) will remain on the record as testimonies and guides to future generations.

In lieu of a conclusion

Although having signed the Paris Agreement on climate change in New York in
April 2016, Turkey is yet to ratify this international accord and enact domestic
legislation. The country does not seem to take climate change seriously. Let alone
reducing emissions, it does not even commit itself to keeping them constant,
insisting on the need for financial support to apply its already weak commitments.

In December 2018, it was a 15-year-old Swedish girl named Greta Thunberg who captured headlines with her impassioned *"J'Accuse!"* plea to world leaders during the UN Climate Change summit in Katowice, Poland. She told the leaders that they were stealing children's futures. *"We need to keep fossil fuels in the ground"* she said, *"If solutions within this system are so difficult to find, maybe we should change the system itself."*[4] Turkey today is nowhere near facing this truth.

Let alone changing the system, even isolated attempts to prevent some of the ecological and social damage caused by the system are engendering violence. In 2017, at least 207 environmental activists were murdered across the world for resisting mines, deforestation, energy companies and agri-business. The killing of environmental activist, Berta Cáceres[5], in 2016 sent shock waves through the world. Although four men were convicted of Cáceres's murder in 2018, concerns of a cover-up and multinational corporate and government complicity in the murder remain. Response to ecological activism is just as menacing in Turkey. Environmental activists Ali Ulvi Büyüknohutçu and Aysin Büyüknohutçu successfully campaigned against a stone quarry and obtained a court order in May 2017 canceling the company's license in the southern town of Finike. Two months later, they were murdered in their home. The suspect confessed to the murder, saying he had been paid by the mine-owner to kill the middle-aged man and woman, and to make it look like a robbery. But before he could further testify in court, he was found dead in his prison cell, allegedly due to suicide. These are not isolated cases. *Global Witness'* annual report aptly titled "At What Cost?" states that government actors such as soldiers or police were directly responsible for at least 53 of the killings of the environmental activists around the world, often fighting against government supported companies to protect their land from 'development'.[6]

The contributions to this book cover a wide range of multifaceted socio-environmental challenges in Turkey while demonstrating a diverse set of disciplinary and methodological approaches to the country's transforming socio-natures. The ecological struggles in Turkey are certainly a part of a global environmental justice movement against neoliberal capitalist valuation and commodification of nature. But against the backdrop of looming climate change, the contamination of water sources in the face of impending droughts and the poisoning of air and soil, the panorama of the socio-ecological transformation in Turkey reveals that what is at stake is the very sustenance of life itself. Spontaneous *'indigenous'* resistance, erupting frequently and gaining ground outside impacted communities, may only grow to represent a challenge to the neoliberal frenzy, especially if and when extreme weather events begin to shake the urban middle classes out of their complacence.

Notes

1 Further reading: Bilgen-Reinart, 2003; Çoban, 2004; Arsel, 2005.
2 Such as the establishment of regional ecological organizations such as the Northern Forests Defense, Ecological Platform of Burhaniye, Artvin Environmental Platform

216   *Epilogue*

etc., and the establishment of Ekoloji Birliği, an umbrella platform for environmental resistance movements, in 2017.
3 The Real Costs of Coal: Muğla, http://costsofcoal.caneurope.org/
4 Greta Thunberg's speech at COP24 plenary session on 12th December 2018, https://www.youtube.com/watch?v=Z1znxp8b65E
5 2015 Goldman Environmental Prize recipient, indigenous environmental leader from Honduras.
6 www.globalwitness.org/fr/campaigns/environmental-activists/at-what-cost/

References

Akdemir, Ö. (2011). *Kuyudaki Taş: Alman Vakıfları ve Bergama Gerçeği*. Istanbul: Evrensel Basım Yayın.

Arsel, M. (2005). The Bergama imbroglio. In F. Adaman, ed., *Environmentalism in Turkey: Between Democracy and Development*, 1st ed. London: Routledge, pp. 263–275.

Bilgen-Reinart, Ü. (2003). *Biz Toprağı Bilirik! Bergama Köylüleri Anlatıyor*. Istanbul: Metis Yayınları.

Çoban, A. (2004). Community-Based Ecological Resistance: the Bergama Movement in Turkey. *Environmental Politics*, 13(2), pp. 438–460.

Northern Forests Defense (2015). The Third Airport Project Vis-a-Vis Life, Nature, Environment, People and Law. Available at: http://bit.ly/NFDReport (accessed on May 20, 2019)

Sakaryalı, M. (2011). *Kışladağ'dan mektup var: Su perisine mektuplar*. Istanbul: Yeni İnsan Yayınevi.

Index